Real Algebraic Geometry

T0280345

UNITEXT – La Matematica per il 3+2

Volume 66

For further volumes:
http://www.springer.com/series/5418

Vladimir I. Arnold

Real Algebraic Geometry

Translated by Gerald G. Gould and David Kramer

 Springer

Vladimir I. Arnold
Steklov Mathematical Institute
Russian Academy of Sciences
Moscow
Russia

Editors
Ilia Itenberg
Université Pierre et Marie Curie and
Institut universitaire de France
Institut de Mathématiques de Jussieu
Paris
France

Viatcheslav Kharlamov
CNRS - IRMA
University of Strasbourg
Strasbourg
France

Eugenii I. Shustin
Fac. Exact Sciences,
School of Mathematical
Sciences
University of Tel Aviv
Tel Aviv
Israel

Translators
Gerald G. Gould
Cardiff University
School of Mathematics
Cardiff
United Kingdom

David Kramer
Plainfleld, MA
USA

Originally published as "Veshchestvennaya algebraicheskaya geometriya," MCCME (c) 2009

ISSN 2038-5722   ISSN 2038-5757 (electronic)
ISBN 978-3-642-36242-2   ISBN 978-3-642-36243-9 (eBook)
DOI 10.1007/978-3-642-36243-9
Springer Heidelberg New York Dordrecht London

Library of Congress Control Number: 2013933709

© Springer-Verlag Berlin Heidelberg 2013
This work is subject to copyright. All rights are reserved by the Publisher, whether the whole or part of the material is concerned, specifically the rights of translation, reprinting, reuse of illustrations, recitation, broadcasting, reproduction on microfilms or in any other physical way, and transmission or information storage and retrieval, electronic adaptation, computer software, or by similar or dissimilar methodology now known or hereafter developed. Exempted from this legal reservation are brief excerpts in connection with reviews or scholarly analysis or material supplied specifically for the purpose of being entered and executed on a computer system, for exclusive use by the purchaser of the work. Duplication of this publication or parts thereof is permitted only under the provisions of the Copyright Law of the Publisher's location, in its current version, and permission for use must always be obtained from Springer. Permissions for use may be obtained through RightsLink at the Copyright Clearance Center. Violations are liable to prosecution under the respective Copyright Law.
The use of general descriptive names, registered names, trademarks, service marks, etc. in this publication does not imply, even in the absence of a specific statement, that such names are exempt from the relevant protective laws and regulations and therefore free for general use.
While the advice and information in this book are believed to be true and accurate at the date of publication, neither the authors nor the editors nor the publisher can accept any legal responsibility for any errors or omissions that may be made. The publisher makes no warranty, express or implied, with respect to the material contained herein.

Cover design: Beatrice 8, Milano

Printed on acid-free paper

Springer is part of Springer Science+Business Media (www.springer.com)

Publisher's Foreword

The preparations for this English-language edition of Vladimir Arnold's *Real Algebraic Geometry* began in the year 2009. With the sad and unexpected death of Arnold on June 3, 2010, publishing this book became a much more difficult task, and it was only with the tireless support and work of several of our collaborators and partners that publication became possible.

Vladimir Arnold had read a first part of the translation prepared by Gerald Gould, and in characteristic fashion, he had sent us, on November 1, 2009, a handwritten letter with his corrections and remarks. He never saw the first complete draft of the translation.

We then asked Ilia Itenberg, Viatcheslav Kharlamov, and Eugenii Shustin to act as editors for the book. We are very grateful to them and thank them for their excellent work. They not only read and checked all the mathematical details of the English edition, but supplied a set of end-notes and comments on the history of Gudkov's conjecture, which we hope will be appreciated by the readers of this book.

We thank Boris Khesin for drawing our attention to the original Russian edition of the book and for his help in enlisting the team of editors. Moreover, we thank Elionora Arnold for handling the formalities with the publishing agreement, thus clearing the legal path to publication.

Special thanks go to Ivan Yashchenko and Yuri Torkhov, of the Moscow Center for Continuous Mathematical Education. They readily agreed to Vladimir Arnold's request to include in this English edition an article of his that had originally been published in one of their journals; this now appears in Appendix A.

Reliably as always, David Kramer did the language editing of the text, and we thank him for his effort.

Finally, we thank our colleague Francesca Bonadei for arranging the publication of this book in the UNITEXT series.

Martin Peters

Foreword

The book you have opened is not a systematic treatment of real algebraic geometry. It was designed as lecture notes destined for high-school students.

The various topics that are discussed in these notes have three features in common: they are about geometry; they treat objects that can be described by algebraic equations; and in most cases, the objects and equations are real. This certainly is one of the reasons for the title "Real Algebraic Geometry." We conjecture that another—and perhaps even deeper—reason for the title is a certain polemic opposing *real* algebraic geometry, in the sense of the algebraic geometry that is closest to *reality*, to other kinds of algebraic geometry. Such a polemic is very much in Arnold's spirit (cf. pp. 37–38).

The reader should not expect to find in this book a formal answer to questions such as, "What is real algebraic geometry?" and "What are its subject matter, main problems, and achievements?" To answer such questions is not the aim of this book (in fact, real algebraic geometry, as a mathematical field, is still in search of its identity).

Arnold has addressed the book to an open-minded reader who is ready to travel with him in a labyrinth of solved and open problems whose formulations are accessible to everyone with a basic knowledge of mathematics, but whose solutions require a certain ingenuity. The text is written in Arnold's brilliant style. Explanations are transparent; the concepts and results are illustrated by numerous examples, interesting digressions to other areas of mathematics and to physics, historical facts, and anecdotes. At the same time, the text is not very polished. Perhaps Arnold wanted to preserve the lecture style of these notes (which certainly made the translator's task difficult, especially since he attempted to replicate the original word for word as much as possible).

We found it necessary to provide comments at certain points in the text, and these appear as numbered endnotes. Finally, we are grateful to G. M. Polotovsky for his help in collecting the materials related to the history of Gudkov's conjecture (see "Editors' Comments on Gudkov's Conjecture," which appear at the end of the book).

We are sure that the reader will enjoy this unordinary book.

The editors

Contents

Chapter 1
Introduction

This book is concerned with one of the most fundamental questions of mathematics: the relationship between algebraic formulas and geometric images.

At one of the first international mathematical congresses (in Paris in 1900), Hilbert stated a special case of this question in the form of his 16th problem (from his list of 23 problems left over from the nineteenth century as a legacy for the twentieth century).

In spite of the simplicity and importance of this problem (including its numerous applications), it remains unsolved to this day (although, as you will now see, many remarkable results have been discovered).

Let f be a polynomial (with real coefficients) of degree n in two variables x and y. Hilbert's question consists in investigating what topological structure an algebraic curve can have if that curve is defined in the Euclidean plane with Cartesian coordinates x and y by the equation[1]

$$f(x,y) = 0.$$

Example. If $n = 1$, then this equation defines a straight line, and all straight lines have the same topological structure.

If $n = 2$, then, as you know, the equation can define, for example, a circle

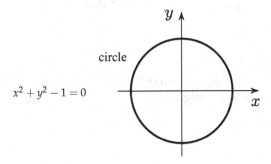

circle

$x^2 + y^2 - 1 = 0$

a hyperbola

V.I. Arnold, *Real Algebraic Geometry*, UNITEXT – La Matematica per il 3+2 66,
DOI 10.1007/978-3-642-36243-9_1, © Springer-Verlag Berlin Heidelberg 2013

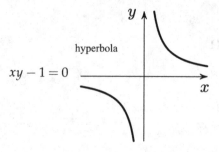

hyperbola

$$xy - 1 = 0$$

or a hyperbola in another form

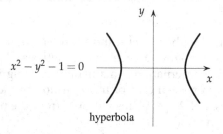

$$x^2 - y^2 - 1 = 0$$

hyperbola

The circle and hyperbola are topologically inequivalent: the circle is connected, while each hyperbola consists of two connected components (called branches) going off to infinity (along the "asymptotes" $\{x = 0\}$ and $\{y = 0\}$ for the first hyperbola, and $\{y = x\}$ and $\{y = -x\}$ for the second).

An equation of the second degree can also define a parabola

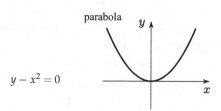

parabola

$$y - x^2 = 0$$

(which differs topologically from a circle and a hyperbola). Indeed, it is topologically equivalent to a straight line.

Chapter 2
Geometry of Conic Sections

The ancients knew quite well that the only curves that can be described by a second-degree equation are the ellipse, the parabola, and the hyperbola; in addition, there are the singular cases of a pair of intersecting lines

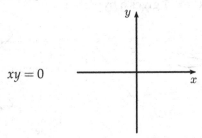

$$xy = 0$$

and a pair of parallel lines

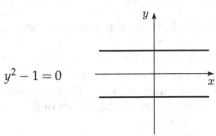

$$y^2 - 1 = 0$$

which in an even more singular case can coalesce to a single line

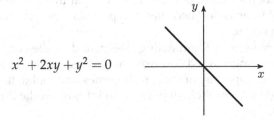

$$x^2 + 2xy + y^2 = 0$$

V.I. Arnold, *Real Algebraic Geometry*, UNITEXT – La Matematica per il 3+2 66, DOI 10.1007/978-3-642-36243-9_2, © Springer-Verlag Berlin Heidelberg 2013 (and in the most singular case of all, in which the polynomial f is the polynomial that is identically zero, the equation $f(x,y) = 0$ describes the entire plane).[2]

The ancients called second-degree curves conic sections; this is because they are obtained by taking the intersection of various planes and the cone

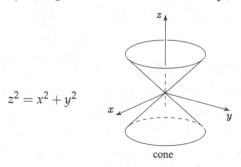

$$z^2 = x^2 + y^2$$

cone

For example, the section with the plane $z = 1$ gives a circle, while the section with the plane $x = 1$ gives a hyperbola

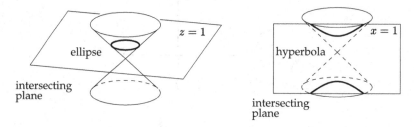

Problem. Which section of the cone by a plane gives a parabola?

Solution. We shall tilt the horizontal plane $z = 1$ continuously towards the vertical position of the plane $x = 1$.

To begin with, the circle distorts (into an ellipse) but remains a closed curve of intersection. But as the angle of tilt is increased, the figure becomes ever more drawn out. At the moment when the tilted plane becomes parallel to one of the lines forming the cone, the length of this intersection goes off to infinity—and this is a parabola.

On tilting the plane yet further (very slightly), we observe that it starts to intersect the lower half of the cone ($z < 0$) as well (somewhere a very long way off), and not just the upper half: beginning at this moment of parabolicity, the line of intersection becomes a hyperbola (and just before this moment, it was an ellipse).

All this can be seen by illuminating the ground by the cone of rays from a torch. If it is strongly tilted towards the ground, then the light patch is elliptic. When the upper ray of the cone becomes horizontal, the illuminated part is bounded by a parabola: it goes off to infinity (in the direction of the horizontal ray).

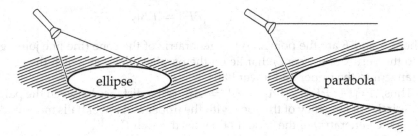

If, on the other hand, some of the rays of the cone of light go upwards, then the boundary of the light patch is hyperbolic (inside is contained an entire sector of light rays of different directions):

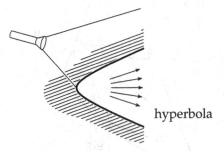

Definition. An *ellipse* is the locus of points P in the plane such that the sum of the distances from P to two fixed points F_1 and F_2 (called the foci of the ellipse) is constant.

$$|PF_1| + |PF_2| = \text{const}$$

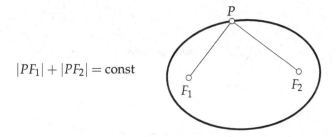

Problem. Prove that the section of a circular cone by a plane whose inclination from the axis of the cone exceeds the inclination of the axis from the lines that generate the cone is an ellipse.

Solution. We inscribe two balls in the cone: one above, and the other below the plane. We lower the upper inscribed ball (and increase its size) until the plane of intersection is tangent to it. Denote the point of tangency by F_1. Similarly, we raise the lower inscribed ball (decreasing its size) until the plane of intersection is tangent to it. Denote the point of tangency by F_2.

Now, two tangents to a ball that are drawn from the same point have equal lengths. Therefore, for every point P on the line of intersection of the plane with the surface of the cone we have the equalities

$$|PF_1| = |PA|, \quad |PF_2| = |PB|,$$

where A and B are the points on the generatrix of the cone (the line joining P to the vertex of the cone) that lie on the parallel circles at which the cone is tangent to the upper and lower balls.

Thus, $|PF_1| + |PF_2| = |PA| + |PB| = |AB|$ is the distance between the parallel circles of tangency of the cone with the upper and lower balls measured along the generatrix of the cone that passes through P.

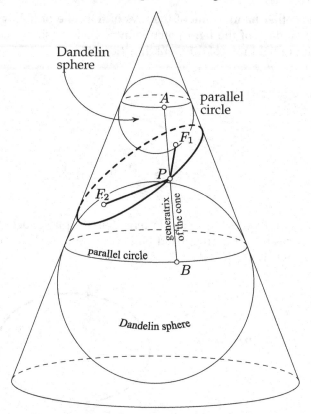

But this distance along any generatrix is always the same, since on rotating the cone around its vertical axis, both spheres tangent to the cone get taken into themselves. Thus, $|PF_1| + |PF_2| = $ const, that is, the line of intersection of the cone with the plane is an ellipse (with foci F_1 and F_2).

The spheres used in this proof are called *Dandelin spheres* (in honor of the mathematician who discovered the above proof).

Problem. Prove that the section of a cylinder by a plane cutting its axis (at a single point) is an ellipse and that all ellipses can be obtained from cylinders in this way.

Hint. Consider the Dandelin spheres inscribed in the cylinder.

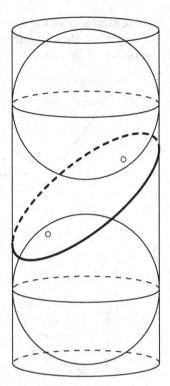

Definition. A *hyperbola* is the locus of points P in the plane such that the difference of the distances from P to two fixed points F_1 and F_2 (called the *foci* of the hyperbola) is constant in absolute value:

$$||PF_1| - |PF_2|| = \text{const}$$

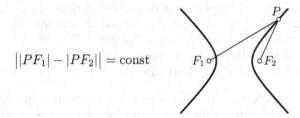

Problem. Prove that the section of the surface of a cone by a plane intersecting both halves of the cone is a hyperbola.

Hint. Construct the corresponding Dandelin spheres, one in the upper half of the cone and the other in the lower half, and both tangent to the cone along entire parallel circles.

The distances from P to the upper sphere along both tangents PF_1 and PA are equal. Similarly, the distances from P to the lower sphere along both tangents PF_2 and PB are equal. Therefore the difference between the lengths of the tangents to both spheres is equal to the distance between the points

A and B of the parallel circles of tangency of the spheres and the cone along the generatrix PAB of the cone joining these points of tangency (here, PAB is a line passing through the vertex of the cone).

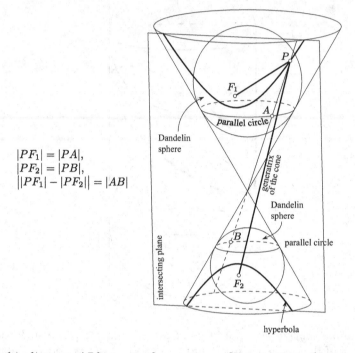

$$|PF_1| = |PA|,$$
$$|PF_2| = |PB|,$$
$$||PF_1| - |PF_2|| = |AB|$$

But this distance AB between the corresponding points on the parallel circles does not depend on the choice of the point P, since on rotating the cone around its axis, both spheres (and their parallel circles of tangency with the cone) are taken into themselves, and any generatrix of the cone is taken to any another generatrix (under a suitable rotation). Thus, the line of intersection of the cone with the plane is a hyperbola with foci at the points F_1, F_2 of tangency of the plane with the Dandelin spheres.

Definition. A *parabola* is the locus of points in the plane equidistant from a line D (called the *directrix of the parabola*) and some point of the plane (called the *focus of the parabola* and denoted by F in the diagram) not lying on the directrix.

$$|PF| = |PA|$$

Problem. Prove that a plane parallel to a generatrix of a cone (and not passing through its vertex) intersects the cone in a parabola.

Hint. Inscribe the Dandelin sphere tangent to both the plane and the cone into the part of the cone intersected by the plane that is adjacent to the vertex of the cone.

$$|PA| = |PB| = |PF|$$

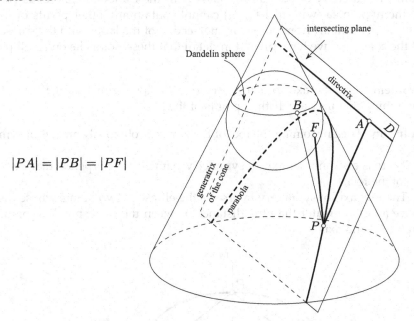

We denote by F the point of tangency of the sphere with the plane. For any point P on the line of intersection of the plane with the surface of the cone, the length PF of the tangent to the sphere is equal to the length PB of the generatrix of the cone tangent to the sphere and joining P to the vertex of the cone.

Then B lies on the parallel circle along which the sphere is tangent to the surface of the cone. The (horizontal) plane of this circle intersects the original cutting plane along the (horizontal) line D.

The length $|PB|$ (of the segment of the generatrix of the cone between P and the circle of tangency) is equal to the length of the perpendicular PA to the line D, since these two are inclined equally relative to this horizontal plane (because the original cutting plane is parallel to one of the generatrices; and all generatrices are equally inclined to the horizontal plane of the parallel).

Thus, the distance $|PF|$ from P to F (the point of contact of the Dandelin sphere with the cutting plane) is equal to the distance $|PA|$ from P to the directrix D (along which the plane of the parallel circle intersects the cutting plane). Hence the line of intersection of the surface of the cone with a plane parallel to one of its generatrices is a parabola (with focus F and directrix D).

Remark. It is readily verified that at other points of the cutting plane, the distances to F and to D are different, since the parabola with focus F and directrix D contains no other points apart from the points of intersection of the cutting plane with the surface of the cone.

In the same way it can be shown that the ellipse (with foci F_1, F_2 at the points of tangency of the cutting plane with the Dandelin sphere) as well as the hyperbola (with foci F_1, F_2) cannot contain any other points of the cutting plane apart from points of intersection of the latter with the surface of the cone (and we have proved in detail that these points lie on an ellipse or hyperbola).

Problem. Find the axes of symmetry of an ellipse with foci at F_1 and F_2 (assuming that they are distinct points of the plane).

Solution. The line joining the foci of the ellipse is obviously an axis of symmetry.

There is also another axis of symmetry, namely, the perpendicular bisector of the foci.

The first axis of symmetry intersects the ellipse at two points whose distance apart is greater than the distance between the corresponding points on the second axis.

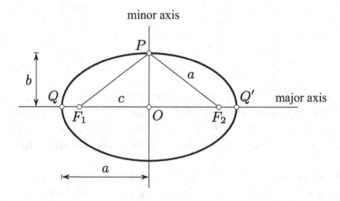

For this reason, the axis of symmetry passing through the foci is called the *major axis* of the ellipse, while the axis of symmetry perpendicular to it is called the *minor axis* of the ellipse.

The half-length of the major axis of the ellipse is denoted by the letter a (and is called the "semimajor axis," although it would be better to call it the "length of the semimajor axis").

The half-length of the minor axis of the ellipse is denoted by the letter b (and is called the "semiminor axis").

Problem. Find the sum of the distances to both foci for points of an ellipse with semiaxes a and b.

Solution. For the point Q of intersection of the ellipse with the major axis, the sum of the distances $|QF_1| + |QF_2| = |QF_2| + |F_2Q'|$ is equal to $2a$, and therefore the sum of the distances is equal to the length $2a$ of the major axis for any point of the ellipse.

Problem. Prove that if a circle is compressed with respect to one of its coordinate axes, then the circle is converted into an ellipse.

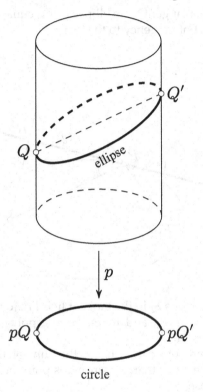

Solution. The projection p of a cylinder onto its circular base uniformly compresses a cutting plane of the cylinder (along the direction of the line QQ' of greatest inclination to the base).

Corollary 1. In a system of Cartesian coordinates in the plane of an ellipse with origin at the point of intersection of the axes of symmetry, with the x-axis directed along the major axis and the y-axis along the minor axis, the equation of the ellipse has the form

$$\frac{x^2}{a^2} + \frac{y^2}{b^2} = 1.$$

Indeed, the equation of a circle of radius 1 in the plane with Cartesian coordinates $X = x/a$, $Y = y/b$ has the form $X^2 + Y^2 = 1$.

Corollary 2. A point of the Euclidean plane with Cartesian coordinates

$$(x = a\cos\varphi, \ y = b\sin\varphi)$$

ranging from $\varphi = 0$ to $\varphi = 2\pi$ describes an ellipse with semiaxes a and b.

Indeed, the point $(X = \cos\varphi, \ Y = \sin\varphi)$ describes a circle of radius 1.

Problem. Prove that a tangent to an ellipse forms equal angles with the two lines joining the point of tangency to the foci.

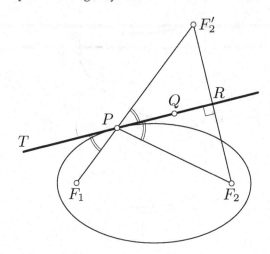

Solution. Reflect the focus F_2 in the tangent line TP. Join the reflected point F_2' to P. Then $|F_2P| = |F_2'P|$, and therefore, the polygonal line F_1PF_2' has length $|F_1P| + |F_2P| = 2a$.

For any other point Q of the line T, the sum of the distances to the foci $|QF_1| + |QF_2|$ is greater than $2a$, since this point lies outside the region bounded by the ellipse.

Therefore, the polygonal line F_1PF_2' is shorter than the polygonal line F_1QF_2', and consequently, this first polygonal line is the shortest path from F_1 to F_2'.

This means that it is a straight line, and the (vertically opposite) angles $\angle TPF_1$ and $\angle QPF_2'$ are equal. But the angles $\angle QPF_2'$ and $\angle QPF_2$ are equal (because F_2' is the mirror image of F_2). Hence the angles $\angle TPF_1$ and $\angle QPF_2$ are equal, as required.

Problem. Find the locus of the feet R of the perpendiculars drawn from the focus F_2 of an ellipse to all of its tangents.

Solution. If $F_1 = F_2$ (that is, $a = b$ and the ellipse is a circle), then the circle itself will be the locus of the feet R of the perpendiculars.

If $2a = |F_1F_2|$, so that the ellipse degenerates into a line segment joining the foci, then all lines passing through the focus F_1 play the role of the tangents TQ.

The foot of the perpendicular F_2R is then the vertex of a right triangle with hypotenuse F_1F_2. Such a vertex lies on a circle with diameter F_1F_2. So in this (singular) case, the locus of the feet R of the perpendiculars turns out to be a circle.

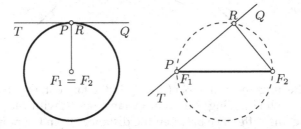

We now consider the general case. First we find the locus of the reflected foci F_2' of the previous problem.

Since $|F_1F_2'| = |F_1P| + |PF_2| = 2a$, this locus of F_2' is also a circle (with centre at the focus F_1 and radius $2a$).

The foot R of the perpendicular is obtained from F_2' as the midpoint of the segment joining F_2' to the focus F_2. Therefore the locus of R is obtained from the locus (which is a circle) of F_2' as the result of the (double) contraction to the point F_2.

Under this contraction the circle is converted to a circle of half the size (of radius a and with centre at a point O lying halfway along the segment F_1F_2, that is, at the centre of the original ellipse). This circle is tangent to the ellipse at the ends of the major axis (as was expected at the very outset).

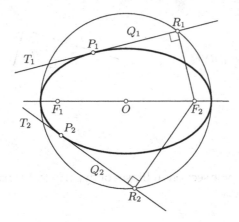

Problem. Prove that a tangent to a hyperbola forms equal angles with the lines joining the point of tangency P with the foci F_1, F_2 of this hyperbola.

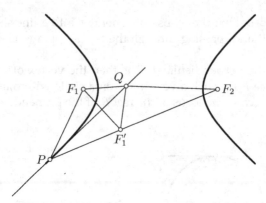

Hint. By reflecting one of the foci (say, F_1) in the tangent, we obtain a point F_1' lying on the line joining the point of tangency P with the other focus F_2; otherwise, the difference between the distances from the point P on the hyperbola to the foci would be (in absolute value) smaller than the distance $|F_1'F_2|$; this is because the difference in the lengths of two sides of a triangle is less than the length of the third side (provided that the triangle is not degenerate):

$$\left||QF_1| - |QF_2|\right| = \left||QF_1'| - |QF_2|\right| < |F_1'F_2| \quad \text{if } Q \neq P.$$

Problem. Prove that an ellipse and a hyperbola with common foci F_1 and F_2 are orthogonal (at their points of intersection).

Hint. The bisectors of adjacent angles are perpendicular.

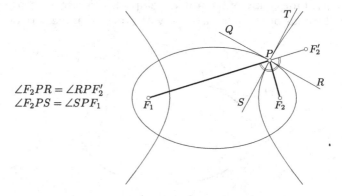

$$\angle F_2PR = \angle RPF_2'$$
$$\angle F_2PS = \angle SPF_1$$

Definition. Two conic sections (ellipses or hyperbolas) are said to be *confocal* if their foci coincide.

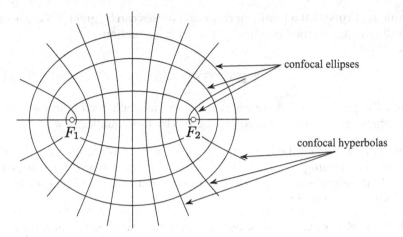

confocal ellipses

confocal hyperbolas

Problem. Consider the Joukowski mapping $f : (\mathbb{C} \setminus 0) \to (\mathbb{C} \setminus 0)$ taking each point $z = x + iy \neq 0$ to the point $w = z + \frac{1}{z}$.

Prove that the circles $|z| = r$ in the z-plane are taken to (confocal) ellipses in the w-plane.

Hint. Denoting the modulus of z by r and its argument by φ, we see that

$$z = r(\cos \varphi + i \sin \varphi), \quad \frac{1}{z} = r^{-1}(\cos \varphi - i \sin \varphi).$$

Therefore at points of the circle where $|z| = r$, we have

$$w = \left(r + \frac{1}{r} \right) \cos \varphi + i \left(r - \frac{1}{r} \right) \sin \varphi,$$

so that the complex number $w = u + iv$ has the form

$$u = a \cos \varphi, \quad v = b \sin \varphi,$$

where $a = r + \frac{1}{r}$, $b = r - \frac{1}{r}$.

Hence, as φ ranges from 0 to 2π, the point $w(\varphi)$ describes an ellipse with major axis a directed along the u-axis, with minor axis b and with centre $w = 0$.

The distance from the centre to the foci in the right triangle on p. 10 is $c = \sqrt{a^2 - b^2}$ by the Pythagorean theorem. But

$$a^2 - b^2 = \left(r + \frac{1}{r} \right)^2 - \left(r - \frac{1}{r} \right)^2 = 4,$$

so that $c = 2$ is independent of r. Therefore the ellipses corresponding to different values of r are confocal: their foci $w = \pm 2$ do not depend on r.

Problem. Prove that a family of confocal ellipses can be given (in Cartesian coordinates x, y of the Euclidean plane) by the equations

$$\frac{x^2}{a^2+\lambda} + \frac{y^2}{b^2+\lambda} = 1, \tag{1}$$

where the parameter λ determines which actual conic section in the family of confocal ellipses (or hyperbolas) is given by this equation.

Problem. Prove that the lines $\arg \varphi = \text{const}$ in the z-plane are taken by the Joukowski mapping $w = f(z)$ into hyperbolas in the w-plane that are confocal with the ellipses corresponding to the family of circles $|z| = \text{const}$ under the Joukowski mapping.

Solution. Multiplication by a nonzero complex number preserves angles. Therefore, two curves in the z-plane forming an angle α at their point of intersection are taken by the mapping $z \mapsto f(z)$ to two curves intersecting at the same angle.

The lines $\arg z = \text{const}$ are orthogonal to the circles $|z| = \text{const}$; therefore, the images of these lines are orthogonal to the confocal ellipses that are the transforms of the circles.

But the hyperbolas confocal with these ellipses are orthogonal to them, as we saw above. Hence the images of the lines $\arg z = \text{const}$ are these hyperbolas.

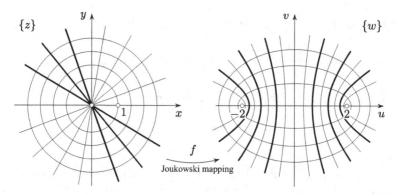

Joukowski mapping

Here the circle $|z| = 1$ is taken to the line segment joining the focal points ± 2, covering it twice. The exterior of this circle in the z-plane is an annulus, and both the exterior and the interior are (smoothly) mapped to the complement (with respect to the w-plane) of the segment joining the foci.

Through each point (x,y) of the plane there pass one ellipse and one hyperbola of the confocal family. Therefore, the quadratic equation (1) in the variable λ (where $a^2 \neq b^2$) has at each (x,y) two real roots λ_1 and λ_2. The functions $\lambda_1(x,y)$ and $\lambda_2(x,y)$ form a system of (orthogonal) *elliptic coordinates* in the (x,y)-plane.

Problem. Investigate the system of elliptic coordinates $(\lambda_1, \lambda_2, \lambda_3)$ in 3-dimensional Euclidean space with Cartesian coordinates (x, y, z) given by the equation

$$\frac{x^2}{a^2 + \lambda} + \frac{y^2}{b^2 + \lambda} + \frac{z^2}{c^2 + \lambda} = 1$$

(where $a^2 < b^2 < c^2$). Prove that of the three surfaces $\lambda_1(x, y, z) = \text{const}$, $\lambda_2(x, y, z) = \text{const}$, $\lambda_3(x, y, z) = \text{const}$, one is an ellipsoid, the second is a hyperboloid of one sheet, and the third is a hyperboloid of two sheets.

Prove that the surfaces of these three types intersect orthogonally in pairs (at each point of Euclidean space with Cartesian coordinates x, y, z).

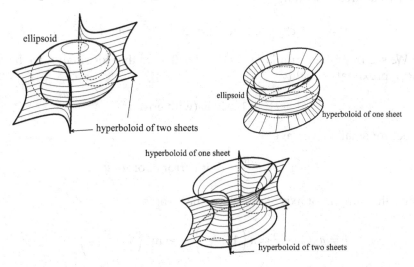

Draw the lines of the system of elliptic coordinates on the surface of the ellipsoid and the hyperboloid of one sheet. Prove the orthogonality of these coordinate curves and investigate the singularities of this system of coordinates.

Definition. For an ellipse with semiaxes $a > b$, the distance from the foci to the centre of the ellipse (the point of intersection of the axes) is denoted by c, and the ratio c/a is called the *eccentricity*.

Problem. Find the length of the semiminor axis of an ellipse whose semimajor axis has length $a = 1\,\text{m}$ and whose eccentricity e is equal to 0.1 (the foci of which are 10 cm from the centre).

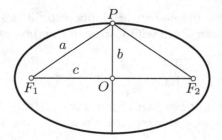

Solution. In the triangle F_1OP, we have $|F_1P| = a$, $|OP| = b$, $|OF_1| = c$. By the Pythagorean theorem,

$$b^2 = a^2 - c^2, \quad \text{so that} \quad b = \sqrt{a^2 - c^2} = a\sqrt{1 - e^2}.$$

We see from the formula $(1 + t)^2 = 1 + 2t + t^2$ that for small t, we have the approximate equality

$$(1 + t)^2 \approx 1 + 2t \quad \text{(with error } t^2\text{)};$$

hence for small u, we have

$$\sqrt{1 + u} \approx 1 + \frac{u}{2} \quad \text{(with error of order } u^2\text{)}.$$

Thus, the semiminor axis of our ellipse has length

$$b = a\sqrt{1 - e^2} \approx a\left(1 - \frac{e^2}{2}\right) = (1\,\text{m})\left(1 - \frac{0.1^2}{2}\right),$$

so that the semiminor axis is smaller than the semimajor axis by approximately $1/200$ of its length (that is, by about a half-centimetre). Thus, an ellipse with eccentricity $e = 0.1$ is practically indistinguishable from a circle (even though its foci are quite a long way from the centre O).

Chapter 3
The Physics of Conic Sections and Ellipsoids

Kepler discovered the laws of motion of the planets by investigating the observations of the planet Mars that were carried out over many years by his teacher Tycho Brahe.

The eccentricity of this elliptic orbit is $e \approx 0.1$. Therefore, Kepler initially came to the conclusion that Mars moves in a circle displaced from the Sun so that the Sun is located at a distance from the centre of this circle equal to one-tenth of the circle's radius.

It was only later that Kepler, using the immense accuracy of Tycho Brahe's observations, noted that the orbit is not quite circular (and arrived at the conclusion that it is an ellipse).

Thus the law that the planets move in elliptical orbits (Kepler's first law) was discovered by Kepler only thanks to his good understanding of the geometry of conic sections; this understanding included the solution of the last problem of the previous chapter:

$$a - b \approx \frac{e^2}{2}a. \tag{2}$$

Interestingly, Tycho Brahe's observatory produced all the observations used by Kepler without any telescopes. The stars and the planets were observed through a small aperture from points of an arc of radius approximately 20 m with centre in this aperture. Here the displacement of a celestial body by an angle α of one degree corresponds to a shift AB of the point of observation by three and a half centimetres, so that the measurement of angles of one minute (1/60 of a degree) was entirely feasible.

V.I. Arnold, *Real Algebraic Geometry*, UNITEXT – La Matematica per il 3+2 66,
DOI 10.1007/978-3-642-36243-9_3, © Springer-Verlag Berlin Heidelberg 2013

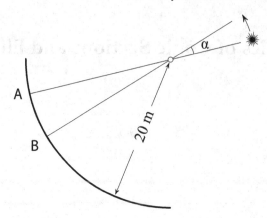

It was only such high accuracy of observation that enabled Kepler to discover his laws.

A hundred years later, Newton sent his pupil Halley to Tycho Brahe's observatory with the assignment of demonstrating there that telescopic measurements can also achieve the same accuracy: the astronomers at Tycho Brahe's observatory did not trust this, since the lengths of the telescopes at that time reached barely 1 metre, so that the variable AB corresponding to an angle $\alpha = 1°$ was less than a couple of millimetres.

The solution of this technical problem of accurate measurement of angles required considerable effort. Here I shall not describe the systems of micrometer callipers with the aid of which the problem was solved (and without which neither telescopic measurement nor rocket launching would have been possible).

Here is another technical application of the same idea, which leads to formula (2). In the construction of the first jet aeroplanes, the hot jet of the engine burned the aeroplane's tailpiece. The designers suggested that the engines be turned slightly so as to divert the hot jet from the body of the aircraft.

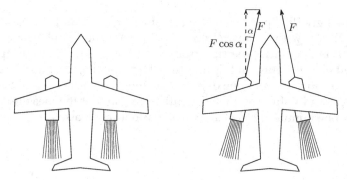

A rotation by an angle α causes the jet to deviate by a distance proportional to α. But as a result of the slight change of direction, the propelling force F pulls the aircraft forward only by its longitudinal component $F \cos \alpha$.

By the Pythagorean theorem, the loss of propelling force is

$$F(1 - \cos\alpha) \approx F\frac{\alpha^2}{2}.$$

For example, for an angle of $3°$ (which already causes a substantial deviation of the jet), the loss is $F(1/20)^2/2$, that is, approximately one eight-hundredth of the propelling force!

Problem. Mr. N returns home, not along a straight line, but by a sinusoidal one. How much longer than the linear distance from home is the route that he traverses?

Most people think that the sinusoidal route is at least double, or at any rate, one-and-a-half times the length of the direct route. But it is in fact only 20% (approximately) longer than the direct route.

This is explained by the same formula

$$\sqrt{1+t^2} \approx 1 + \frac{t^2}{2}$$

(with error of order t^4) for small t. Here t is the tangent of the angle between the curved and straight trajectories.

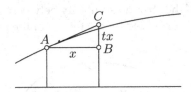

By the Pythagorean theorem, the hypotenuse of the right triangle ABC (which is small for small x) with a small angle (whose tangent is t) at a vertex is given by the formula

$$\sqrt{x^2 + t^2 x^2} = x\sqrt{1+t^2} \approx x\left(1 + \frac{t^2}{2}\right).$$

For example, for $t = 1/3$ (which corresponds to an angle of about $20°$ at the vertex A), the hypotenuse is longer than the leg AB by roughly $(1/3)^2/2$, that is, by roughly 5%.

The angle between Mr. N's sinusoidal path and the straight one is nowhere greater than $45°$ and is considerably less on most of the path; and this is why the sinusoidal path exceeds the direct path by such a small amount.

Problem. Calculate (to four decimal places) the square root of 0.996.

Answer. 0.9980.

Problem. Calculate (with error of order u^2) the quantity $\sqrt[3]{1+u}$, where u is small.

Answer. $1+u/3$.

Problem. Calculate (with error of order u^2) the quantity $(1+u)^n$, where u is small.

For example, for $n = 3, 1/5, -1$ one obtains the very important formulas

$$(1+u)^3 \approx 1 + 3u, \quad \sqrt[5]{1+u} \approx 1 + \frac{u}{5}, \quad \frac{1}{1+u} \approx 1 - u.$$

The general formula, which is true for small $|u|$ with error of order u^2,

$$(1+u)^n \approx 1 + nu,$$

together with its refinements

$$(1+u)^n = 1 + nu + \frac{n(n-1)}{2!}u^2 + \frac{n(n-1)(n-2)}{3!}u^3 + \cdots,$$

is called Newton's binomial formula (because it was known by him both for positive integer exponents as well as for other cases).

We now turn to the study of ellipses with small eccentricity.

Another example of the striking similarity between an ellipse of small eccentricity and a circle is provided by the theory of the propagation of waves in an elliptic region.

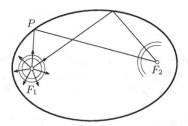

We consider an elliptic cup of water and let fall a drop of water onto one of the foci F_1. From the point F_1, circular waves start to spread along the surface of the water. On reaching the edge, the waves are reflected from it and are then propagated along rays orthogonal to the wave fronts. At the moment of reflection, a ray changes direction in accordance with the law "the angle of incidence equals the angle of reflection." In accordance with the first problem on p. 12, the reflected ray passes through the second focus F_2.

This means that all the reflected rays collect at the second focus. The re-flected waves arrive along these rays at the second focus simultaneously, since all the polygonal lines F_1PF_2 have the same length (equal to $2a$).

Thus, when the waves emerging from F_1 collect at the second focus, their fronts will have the form of circles with centre at F_2, and we observe at this instant a splash at the point F_2.

Suppose now that the ellipse has small eccentricity e. Then it is practically indistinguishable from a circle ($a - b \sim ae^2/2$), even though each focus is noticeably different from the centre ($c \sim ae$).

It follows that the experiment with the elliptical cup described above can also be carried out in an ordinary circular cup in which the surface of the water is bounded by a circle.

More precisely, when we let a drop of water fall on the surface of the water, say at a distance of a third of the radius from the centre, we see how the spherical waves reflected from the edge of the cup again accumulate, creating a noticeable surge at the point F_2 symmetric to F_1 (the starting point of excitation of the wave) with respect to the centre of the cup.

Thus, anyone can demonstrate the property of an ellipse discussed above by letting fall a drop of tea in an ordinary circular cup. The surge at the symmetric point is easily noticed (except, of course, that the tea must not be interfered with at this time; on the contrary, the initial velocities of its particles must be zero).

Problem. Consider in the complex z-plane an ellipse with centre at $z = 0$ and map each point z to the point $w = z^2$.

What is the image of the original ellipse in the complex w-plane?

Solution. The image will be an ellipse with focus at the point $w = 0$.

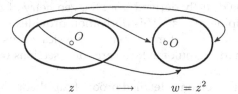

$$z \quad \longrightarrow \quad w = z^2$$

For the proof we use the problem on p. 15. To this end, we again consider a new complex variable t and associate with t the value of the "Joukowski function"

$$z = t + \frac{1}{t}.$$

According to the problem on p. 15, the original ellipse in the z-plane can be obtained from a suitable circle $|t| = r$ in the t-plane (if the foci of the ellipse are located at the points $z = \pm 2$).

We now obtain

$$w = z^2 = t^2 + \frac{1}{t^2} + 2.$$

The point $t^2 + \frac{1}{t^2}$ belongs to an ellipse with foci ± 2 (since $|t^2| = r^2$). The addition of 2 shifts the focus -2 to the point $w = 0$, which proves our assertion (for the original ellipse with foci $z = \pm 2$).

If the foci of the original ellipse are at some other place $\pm F$, then the assertion about the ellipse's image in the w-plane remains true. One can pass from the complex coordinate z to the coordinate $z' = kz$ (with complex coefficient k) by choosing the coefficient k so that $z = 2$ corresponds to $z' = F$ (that is, choosing $k = F/2$). Then, by applying the (proved) assertion to the ellipse with foci at $z = \pm 2$, we obtain it for the ellipse with foci at $z' = \pm F$.

Hooke's law (motion under forces of elasticity) states that the force acting on a displaced point is proportional to the displacement and acts in the direction opposite to that of the displacement:

$$\frac{d^2 \mathbf{x}}{dt^2} = -c\mathbf{x}.$$

It is easy to see that such a point performs harmonic motion about the point $\mathbf{x} = 0$:

$$\mathbf{x}(t) = \mathbf{A}\cos(\omega t) + \mathbf{B}\sin(\omega t) \quad (\text{where } \omega^2 = c).$$

The orbit of this motion is an ellipse with centre at the point 0.

The solution of this problem shows that if one regards the plane of oscillation of the vector x as a complex line and takes the square of the complex number $z = \mathbf{x}$, then one obtains the orbit ($w = z^2$) in the field of universal gravitation (or in the electrostatic Coulomb field), the force of which is directed towards the attracting centre and is inversely proportional to the square of the distance from the attracting centre.

This fact can be directly derived from the geometry of conic sections related by the mapping $w = z^2$. This derivation provides the simplest proof of the most important fact of mathematical physics: orbits in a field of universal gravitation (or a Coulomb field) are conic sections with a focus at the attracting point.

Here I shall not dwell on a detailed proof of this theorem, leaving it to the reader. I merely remark that the mapping $z \mapsto (z^2 = w)$ does not convert the *motion* under the action of a Hooke force into motion under the action of a Newton or Coulomb force, but merely converts the orbits of these motions from one into another; it does not affect the time of the motions along the corresponding (Hooke and Newton or Coulomb) ellipses.

It is interesting to note that for all the possible laws of attraction to the centre by a force whose strength depends only on the distance from the attracting centre, the closed orbits (for the entire range of initial conditions) are obtained only in the two cases considered above, namely, when the force is directly proportional to the first power of the distance from the centre of attraction or inversely proportional to the second power of the distance from the centre of attraction.

Orbits in fields of attraction with forces proportional to distances raised to the power of α (respectively, β) are taken to each other by transformations of the form $z \mapsto (w = z^\gamma)$ if and only if

$$(\alpha + 3)(\beta + 3) = 4,$$

and this strange duality persists even for quantum-mechanical particles. These results can be verified by direct (albeit lengthy) calculations. However, there also exists a geometric proof based on variational "principles of least action" replacing the differential equations of the motion. Some mathematicians regard geometry as though it were "a method of not making mistakes in long calculations." I do not agree with them, although Jean-Jacques Rousseau wisely explained in his *Confessions* how in his childhood he was for a long time unable to remove the parentheses in the algebraic formula $(a + b)^2 = a^2 + 2ab + b^2$ and that the "real proof" came to him only from the following geometric diagram:

ba	b^2
a^2	ab

Confocal families of second-degree algebraic surfaces turn up in a surprising way in the theory of gravitational attraction. The first result in this direction was the following.

Newton's theorem. *A homogeneous sphere attracts points exterior to it with the same force as if its mass were concentrated at the centre. On the other hand, in the interior of the sphere, the resultant force of attraction due to all the points of the sphere is equal to zero.*

The proof of the second assertion is based on the fact that each line intersects the sphere at the same angles at both points of intersection of this line with the sphere.

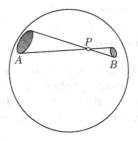

It follows from this that the areas of the opposite parts of the sphere cut out on it by the narrow cone with vertex at the interior point P near points A and B of the sphere are directly proportional to the squares of the distances $|PA|$, $|PB|$.

But the force of attraction is inversely proportional to the square of the distance. Thus the force of attraction at P by the opposite parts are directed from the opposite sides of P and are equal in magnitude. This means that they neutralize each other (giving zero resultant force).

The first part of Newton's theorem also has a simple geometric proof. It is simplest to explain it in hydrodynamic terms: the vector field of attraction towards a point, which is inversely proportional to the square of the distance from it (in Euclidean 3-space), is the velocity field of an incompressible fluid (in the complement of the attracting point).

This follows from the fact that the output of the "sources" of such a velocity field, as well as the field itself, is spherically symmetric. At the same time, the flux of this field through the surface of any sphere with centre at the attracting point ("the amount of liquid passing through this surface in unit time") is the same for any sphere (since the velocity is inversely proportional to the radius of the sphere, while the area of the sphere is directly proportional to the square of the radius).

Hence, a layer between two concentric spheres has zero resultant output of all the sources of our "fluid." Consequently, this output of the sources is equal to zero at each point of the sphere, so that the flowing "fluid" is incompressible.

Following from this remarkable property of the gravitational field of a point is the same incompressibility of the field of attraction for any body at all points outside this body (since these fields are obtained by adding the fields created by the separate particles of the body, and the output of the sources in the sum of the fields is the sum of the outputs of the sources of the fields being added).

It remains to observe that on transferring the particles of a ball to its centre, the flux of the field of attraction of these particles through a sphere with centre at the centre of the ball and with radius greater than the radius of the ball is not altered.

By transferring to the centre all the points of the ball, we obtain a field of attraction that is just as spherically symmetric as the original vector field and will have the same flux as the original field through the surface of any sphere with centre at the centre of the ball.

Hence these two vector fields are the same; this completes the proof of the first part of Newton's theorem.

Problem. A shaft is drilled in a homogeneous ball (along a diameter joining two opposite points of the bounding sphere) and a point mass P starts to fall inside this shaft (with zero initial velocity). Study the motion of this point under the action of the forces of attraction of the particles of the ball.

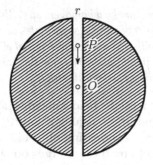

Solution. The point P will perform harmonic motion: its distance x from the centre O of a ball of radius r varies with time in accordance with the formula $x = r\cos(\omega t)$. The calculation of the frequency ω of these oscillations is an interesting supplementary problem; the period of oscillation of the point is related to the one-and-one-half-hour period of rotation around the Earth of a near-Earth satellite with "first cosmic velocity" of around 8 km/sec.

Proof. The parts of the ball that are farther from the centre than the falling mass P do not attract it (by the second part of Newton's theorem). The mass m of the part of the ball at a distance of less than $x = |OP|$ is proportional to the cube of the distance of P from the centre: $m = \text{const} \cdot x^3$.

The total force of attraction of this part is directly proportional to its mass and inversely proportional to the distance of P to the centre: $F = \text{const} \cdot m/x^2 = -cx$.

Hence the gravity field inside the shaft is a "Hooke's-law field," the attracting force to the centre being directly proportional to the distance from the centre.

Since the oscillations due to Hooke's law are harmonic, we see that $x = r\cos(\omega t)$ (from Newton's second equation $d^2x/dt^2 = -cx$).

We now turn to the attraction of aspherical bodies, for example, the Earth, which is close to an ellipsoid whose major axis is greater than its minor axis by 1/300.

In this case, Newton's theorems on the attraction of homogeneous spheres and balls take the following form (usually referred to as "Ivory's theorem").

Theorem. *A homogeneous ellipsoid attracts points situated outside it in the same way as any confocal ellipsoid situated inside it.*

The homogeneous layer between two confocal ellipsoids has no attractive effect whatever on points inside the smaller ellipsoid bounding the layer.

A proof of this theorem by direct calculations is possible, but somewhat lengthy. A geometric proof (along the lines of the proof given above for the spherical case) is also possible. I merely note the following supplement to the theorem given above (which makes its understanding and proof easier).

Theorem. *The force of attraction by a homogeneous ellipsoid is directed along lines (orthogonal to it) of the system of elliptic coordinates (and is orthogonal to the surfaces of ellipsoids confocal with it) outside the original ellipsoid.*

The surfaces of constant potential energy (for this force field of attraction of the ellipsoid) are surfaces of ellipsoids confocal with it.

Since the Coulomb electrostatic field coincides (apart from the sign) with the gravitational force field, these results also provide a description of the distribution of charges on a conducting ellipsoid and a description of the (attracting or repelling) field by these charges outside the ellipsoid. (Inside it the field is zero: "A conductor shields the field," as the physicists say.)

More precisely, all the charge collects on the surface of the ellipsoid, and its density is greater wherever the curvature of the ellipsoid is greater: it is determined by the thickness of the layer between the original ellipsoid and an ellipsoid close to it and confocal with it.

The lines of force of the electric field created outside the ellipsoid represent an elliptic system of coordinates, including the original ellipsoid as a surface of constancy of one of the elliptic coordinates, and these lines are perpendicular to this surface.

All the results about elliptic coordinates in $\mathbb{R}^3$ described above can be carried over to the case $\mathbb{R}^n$; only the size of the "gravitational field of a point" in $\mathbb{R}^n$ is not of the form r^{-2} (the Coulomb and gravitational laws), but of the form $r^{-(n-1)}$ (for incompressibility, the magnitude of a field on a sphere of radius r must be inversely proportional to the n-dimensional "area" of this $(n-1)$-sphere).

For example, for $n = 2$ the "gravitational" field of a point is inversely proportional to the distance from the attracting point.

This can be interpreted physically, for example, as follows. Consider a distribution of masses in Euclidean 3-space with Cartesian coordinates (x,y,z) that is homogeneous along the z-axis (in other words, the distribution is unaltered by rotations about the z-axis).

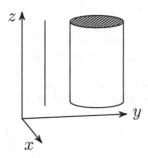

For example, they could be vertical homogeneous rods (of infinite length), infinite vertical cylinders, vertical homogeneous planes, and so on.

The attractive force created by such (vertically standing) masses is invariant with respect to translations along the z-axis (as well as with respect to reflection in a horizontal mirror taking the point (x,y,z) to $(x,y,-z))$. There-fore, such a force is directed horizontally, so that we get a force field in the horizontal (x,y)-plane.

Calculations show that the magnitude of the attractive force due to a ver-tical homogeneous rod (of infinite length) is inversely proportional to the first power of the distance of the attracted point to this rod.

Thus, in the Euclidean horizontal plane we indeed obtain the force field that above we called the "gravitational field in the Euclidean space $\mathbb{R}^2$."

Corollary. *The gravitational field of a homogeneous layer between two confocal ellipses in the Euclidean plane is equal to zero inside the smaller bounding ellipse, while outside the larger bounding ellipse it is directed along hyperbolas confocal with these ellipses.*

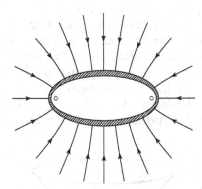

Problem. Prove that the equipotential curves of this field (joining points of equal potential energy of the field) are ellipses confocal with the ellipses bounding the original layer.

Remark. A similar description can also be given for the two-dimensional Coulomb field created by a homogeneous distribution of charges between two confocal hyperbolas, only in this case one must put positive charges along one branch and the same negative charges along the other.

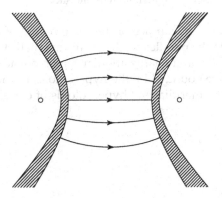

This field is equal to zero in the convex parts of the plane bounded by the arcs of the hyperbola.

In this case the lines of force of the field are the parts of the ellipses confocal with the original hyperbolas that lie between the two charged layers. Here the equipotential curves are hyperbolas that are confocal with the original ones (and lying between both charged layers).

Everything said above generalizes to Euclidean space in other dimensions n. For example, when $n = 3$, one can place opposite charges on the two sheets of a hyperboloid of two sheets extended to the layer between them and a close confocal hyperboloid.

The usual Coulomb field of such a distribution of charges is directed in the region between both sheets of the hyperboloid along the lines of an elliptic system of coordinates orthogonal to this hyperboloid (along which the ellipsoids and the hyperboloids of one sheet of the family intersect).

It is interesting to note that if one starts with a hyperboloid of one sheet in three-dimensional space, then the analogous constructions give one a magnetic field rather than an electrostatic one in the complementary regions.

More precisely, the lines of the elliptic system of coordinates on the hyperboloid of one sheet separate out into two classes: closed "parallels" along which this hyperboloid is intersected by the ellipsoids confocal with it, and the "meridians," which are orthogonal to the parallels and go from infinity to infinity.

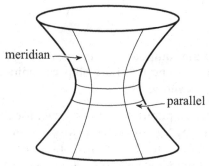

hyperboloid of one sheet

It turns out that on the surface of the hyperboloid of one sheet one can send a current along the parallels such that the magnetic field created by this "solenoid" will vanish in the large annular region outside the hyperboloid, while inside the tube bounded by the hyperboloid, this magnetic field will be directed along the meridians of hyperboloids of one sheet confocal with the original one.

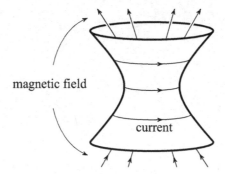

Along the meridians of the original hyperboloid of one sheet one can send a current such that the magnetic field created by this current will vanish inside the tube bounded by the hyperboloid.

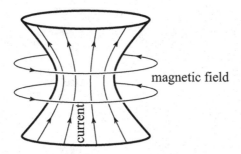

In the large external annular region complementary to this tube, the lines of the magnetic field will be closed curves—they are parallels on the surfaces of hyperboloids of one sheet confocal with the original one and lying in the outer annular region (along which these hyperboloids intersect the ellipsoids confocal with them).

All these assertions (and their multidimensional generalizations) can be proved by direct (albeit lengthy) calculations; but these simple geometrical and physical facts remained unnoticed for a long time (both by mathematicians and by physicists), since their understanding requires simultaneous mastery of the geometry of elliptic coordinates (the foundations of which were laid down by the modern theory of "completely integrable Hamiltonian systems," as well as the equations of the theory of the electromagnetic field, as laid down by the mathematical physicist Maxwell, and in addition, finally, the topology of manifolds (de Rham's theorem)).

The difficult part of the theory arising here has been explained above; the technical details of the calculations are left to the reader.

Chapter 4
Projective Geometry

It is clear to everyone that the branches of a hyperbola "go off to infinity," but these words acquire a precise meaning only in projective geometry, the science that arose in the mathematical examination of rails "converging" a long way off as they go straight to the horizon.

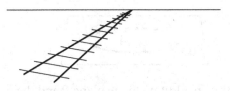

Perspectives of this kind have long been used by artists. For example, in his biography of Paolo Uccello, Vasari relates that this artist (active in the time of Raphael) was so fascinated by drawing perspectives that when his wife called him from his studio to come to bed, he replied "I am coming straightaway—what a beautiful perspective!" (he had in mind the architectural sketches that he was finishing of a town stretching off into the distance).

The term "projective geometry" originates from the projection of one plane onto another produced by rays passing through an "observation point" O.

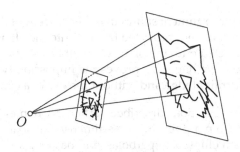

V.I. Arnold, *Real Algebraic Geometry*, UNITEXT – La Matematica per il 3+2 66,
DOI 10.1007/978-3-642-36243-9_4, © Springer-Verlag Berlin Heidelberg 2013

Here is an example of a completely nontrivial fact of projective geometry.

Theorem. It is impossible to construct the centre of a given circle in the plane using a straightedge only, without the use of a compass (to do so using a compass is an easy task).[3]

Proof. Consider an "oblique cone" over the plane at whose base lies the given circle; the vertex O of the cone is orthogonally projected onto the plane to a point distinct from the centre of the given circle.

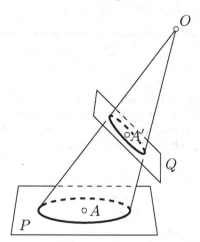

It is easy to see that in addition to the plane P and the planes parallel to it, there is another plane Q that intersects the cone along a circle; the planes P and Q are inclined oppositely with respect to the central axis of the cone. If there existed an algorithm for constructing (using a straightedge) the centre of the circle in the plane P, then some system of lines in this plane (constructed with the help of the given circle) would determine the centre of this circle as the intersection of two lines of this system.

We project this entire system of lines from the centre O onto the plane Q. Then the projected lines would form a system of lines constructed in the same way for the circle in the plane Q. Therefore the projections of the two lines of the system providing the centre would intersect at the centre of the projected circle.

However, a simple argument shows that the centre A of the original circle of the plane P is certainly not projected from O onto the plane Q to the centre A' of the circle projected onto the plane Q; in other words, the points A, A', and O do not lie on one line. This proves the impossibility of constructing the centre using a straightedge and without the use of a compass.

The theory of conic sections described earlier, in Chapter 2, shows that by projecting a circle from a point in three-dimensional space onto a suitable plane one can obtain ellipses, hyperbolas, and parabolas.

Problem. Prove that projections of ellipses, hyperbolas, and parabolas (from various centres and onto various planes) give rise to ellipses, hyperbolas, and parabolas again.

Hint. Investigate whether the composition of two projections is again a projection; if this is the case, then, for example, the projection of an ellipse (from a centre onto a plane) can be obtained as the projection of a circle whose projection is this ellipse.

The mathematical apparatus of projective geometry uses the *projective plane*, obtained from the ordinary plane by adding so-called points at infinity. Here is how it is defined.*

Consider in Euclidean 3-space some point O and all the lines passing through this point. These lines are declared to be the points of the projective plane $\mathbb{R}P^2$. To describe the set of all these lines (that is, the set of all points of the real projective plane $\mathbb{R}P^2$) one can proceed as follows.

We fix in our space some plane P not passing through the chosen point O.

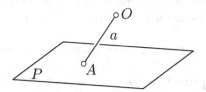

Corresponding to each point A of the plane P there is a line $a = AO$ passing through O. In this way we obtain almost all points a of the projective plane $\mathbb{R}P^2$ from points A of the ordinary Euclidean plane P.

Suppose now that the point A moves along the plane P, going off to infinity in some direction.

The corresponding line $a = OA$ will then rotate around the point O and in the limit (as $A \to \infty$) will tend to a well-defined line a_∞ (parallel to the plane P but passing through O).

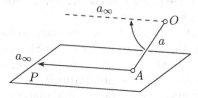

The real projective plane $\mathbb{R}P^2$ is obtained from the "affine part" (which is isomorphic to the plane P and consists of all lines of the form $a = OA$ that are not parallel to the plane P) together with the set of all "points at infinity"

---

* In this connection Goethe wrote, "If you want to understand tending to infinity, you simply need to set off from a point in a finite region in all possible directions."[4]

(such as a_∞), these being all the lines passing through O and parallel to the plane P.

To see this more clearly, consider a sphere of radius 1 with centre at the point O. Rays of the form OA project the plane P onto the "southern hemisphere" (the half turned towards the plane P, in other words, the half "south of the equator").

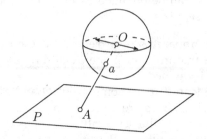

All the points of the projective plane $\mathbb{R}P^2$ are depicted on this sphere by the closed southern hemisphere, including the equator; the latter corresponds to lines parallel to the plane P (and called the "points at infinity" of the projective plane).

However, each such line intersects the equator at two diametrically opposite points, rather than just one point. Therefore, in order to enumerate all the points of the projective plane one should add to the southern hemisphere not the whole equator, but just half of it (taking, for example, the points with longitude $0 \leq \phi < \pi$).

However, it is more convenient to proceed differently and use not just the southern hemisphere but the entire sphere, and then identify diametrically opposite points on it.

The situation here is not all that unusual (for geographic images). For example, on some (old) maps of hemispheres, the Kamchatka Peninsula is pictured twice: both in the Western and the Eastern Hemispheres.

In completely the same way, for the study of neighbourhoods of points at infinity in the projective plane it is convenient to use as a map the whole surface of our sphere (near the equator).

The point is that just as the Earth is constructed in the same way both on the meridian dividing the Western and Eastern Hemispheres and in other places, the projective plane has, in neighbourhoods of the points at infinity, exactly the same properties as in neighbourhoods of the "finite" points of its affine part (which is isomorphic to the original plane P).

Problem. Verify that in the projective plane, a hyperbola turns into a closed connected curve (topologically an ellipse or a circle).

In fact, on the projective plane the hyperbola acquires two points at infinity, $B = B'$ and $D = D'$.

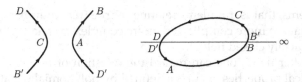

Problem. What does a parabola look like in the projective plane?

Answer. The parabola acquires a unique point at infinity $C = C'$ and turns into a closed smooth curve tangent (at $C = C'$) to the line of points at infinity.

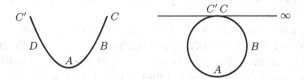

Problem. Draw in a neighbourhood of its point at infinity the cubic parabola given in Cartesian coordinates of the Euclidean plane by the formula $y = x^3$.

Answer. The cubic parabola acquires a unique point at infinity $C = C'$ and turns into a nonsmooth closed curve (with a semicubical cusp):[5]

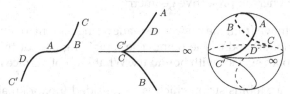

The fact that the ellipse, hyperbola, and parabola become topologically indistinguishable in the projective plane suggests that it is more reasonable to begin the topological classification of algebraic curves of degree n with the study of their completions in the projective plane, where there are fewer classes and the answers are simpler.

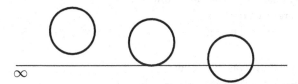

I shall now describe what is currently known about these classifications, although the general answer (for any degree n) is unknown and presents one of the central fundamental questions of mathematics, which remains open in spite of the extensive efforts of numerous celebrated mathematicians. One might think that this question relates to algebraic geometry.

But the contribution of algebraic geometers to this problem is disproportionately small (principally because they are incapable of solving actual (real) problems, that is, problems relating to true (real) numbers, preferring problems that are more complicated, more complex, or related to equations over algebraically closed fields).

On the other hand, one can regard the question of the investigation of the topological properties of curves defined by polynomial equations of the form $f(x,y) = 0$ as related to computer mathematics.

But even here success has been negligible. I shall attempt to describe what is new that has been learned merely by computer calculations in this area. True, there is a remarkable general theorem (the Tarski–Seidenberg theorem) stating that there exists a finite algorithm for solving this kind of question about polynomials.

However, the trouble is that the volume of calculations required for solving problems on the topological classification of all algebraic curves of a given (quite moderate) degree n is so large that the best computers would be unable to cope with them even if they were given the length of time that the universe will exist.[6]

In contrast, mathematicians have obtained a number of perfectly remarkable results in this area (although they do not provide a full solution of the problem).

I begin my account of them with a very simple question posed by one of the first researchers in projective geometry.

Problem. Remove from the projective plane a single point, or even an entire circular neighbourhood of the point. The remaining part of this projective plane is a smooth surface with boundary. What kind of surface is it?

Answer. It is a Möbius strip, which is a one-sided (nonorientable) surface discovered by Möbius in his solution to this problem.[†]

[†] The modern (Mekhmat) poet N. Yu. Ivanova-Filippova described this surface as follows: (see "We are the mathematicians from Lenin Hills," Knizhnik (2007), no. 3, 140–141):

Möbius strip—symbol of mathematics,
Which serves higher wisdom with a crown ...
It is full of unrealized romance:
In it infinity is enfolded by a ring.
⟨...⟩
It seems that Eternity was spread out,
That the password of the Universe was unlocked
And suddenly your aspirations to infinity
Turn you back to the starting point: back to zero.
As against a threshold you stumble against this zero.
But, however prickly the previous path might be,
Choose again (and you make no mistake!)
Road to infinity—Möbius strip!

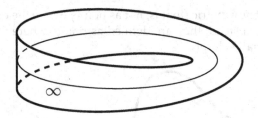

What Möbius did was to investigate a neighbourhood of the line at infinity, that is, the line of points added to the affine plane.[7]

On the sphere doubly covering the projective plane, this line at infinity is depicted by the equator. This means that we are dealing with a neighbourhood of the equator on the surface of the sphere. But in order to get from the sphere to the projective plane, we need to identify diametrically opposite points.

The two parallels (say, $10°$ north and $10°$ south) enclosing the neighbourhood are glued to a single circle under this identification, so that from the (cylindrical) neighbourhood we obtain (after this gluing) something that is not a cylinder.

In fact, how to carry out the gluing is clear from the figure: it suffices to restrict oneself to the part of the equator between the longitudes $0°$ and $180°$ east and glue together just the bounding meridians AB and $A'B'$ (where the points A and A' are diametrically opposite).

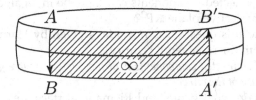

One achieves the usual gluing of a Möbius strip from a rectangular piece of paper $ABA'B'$: and that is how Möbius himself created it.

We see that "infinity is enfolded by a ring" and that "aspirations to infinity (along the projective line of this ring) turn you back to the starting point: back to zero."

Problem (Möbius). Consider a line in the projective plane (for example, the line at infinity or simply the x-axis of the ordinary plane completed by its point at infinity).

Now perturb this line so as to obtain a smooth curve in the projective plane that is not self-intersecting. Can there be fewer than three points of inflection on this curve?

The perturbation of a projective line depicted in the figure below shows that the situation with exactly three points of inflection is possible. Note that the notion of inflection point is a projective one: points of inflection can be defined without any metric (that is, not as points where the curvature of the curve vanishes); instead, they are defined as points where the tangent line touches it unusually closely.

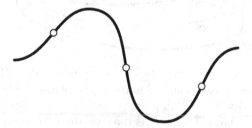

If the difference from the perturbed curve is not too great, then Möbius's assertion about the three points of inflection can be proved (although its proof is not simple).[8]

Problem. Prove that if the perturbed curve is allowed to be self-intersecting, then it may happen that there remains just one point of inflection (if the perturbation is sufficiently large).

Möbius deduced from the nonorientability of the Möbius strip that the number of (nondegenerate) points of inflection is odd; the proof of this theorem is a nice problem.

How many connected components can there be in an algebraic curve of degree n in the projective plane $\mathbb{R}P^2$?

The answer to this question is nowadays given by Harnack's theorem (proved already in the nineteenth century).[9]

The number of connected components is at most $g + 1$, where g is the genus of the Riemann surface of the curve.

We now explain what genus and Riemann surface are together with a proof of the following *Riemann–Hurwitz formula:*[10]

The genus of the Riemann surface of a smooth algebraic curve of degree n in the complex projective plane is equal to

$$g = \frac{(n-1)(n-2)}{2}.$$

Before discussing these two theorems, we apply them to the problem of counting the maximum number of components M that an algebraic curve of degree n may have in the real projective plane $\mathbb{R}P^2$. The previous theorems provide the following values:

n	1	2	3	4	5	6	7	8
g	0	0	1	3	6	10	15	21
M	1	1	2	4	7	11	16	22

For example, the line has one component (for $n = 1$ we have $M = 1$). In the case $n = 2$ we have indicated $M = 1$; indeed, the ellipse, the hyperbola, and the parabola all have one component in the projective plane $\mathbb{R}P^2$.

Problem. Construct a smooth algebraic curve of degree 3 having $M = 2$ connected components in the projective plane $\mathbb{R}P^2$.

Solution. The curve of degree 3 given in the plane with Cartesian coordinates x and y by the equation

$$xy(1 - (x + y)) = 0$$

consists of three lines forming a triangle and is therefore not smooth.

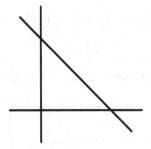

But if we replace 0 on the right-hand side by a small positive number (for example, $1/100$), then the curve becomes smooth.

It changes around the points of intersection of its branches approximately in the same way as the cruciform curve $xy = 0$ is converted to the hyperbola $xy = 1/100$:

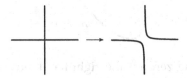

As a result, we obtain the following perturbed curve in the (x,y)-plane

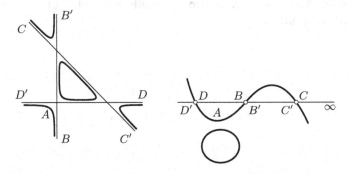

with three points at infinity B, C, D. In the projective plane (right-hand diagram) we have two connected components: one of them is the closed oval in the original affine (x, y)-plane, and the other is formed from the three unbound branches of the affine curve $ABB'CC'DD'$. A full investigation of the topological types of all curves of degrees 3 and 4 was carried out by Newton and Descartes.[11]

Problem. Construct a smooth curve of degree 4 having four connected components in the projective plane $\mathbb{R}P^2$.

Solution. By multiplying two second-degree equations

$$f(x,y) = 0 \quad \text{and} \quad g(x,y) = 0$$

(corresponding to two ellipses), we obtain the fourth-degree equation

$$fg = 0,$$

which defines a nonsmooth curve in the (x, y)-plane if the ellipses intersect:

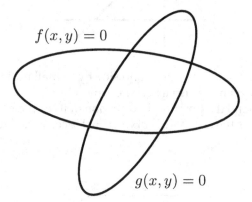

We now replace the zero on the right-hand side of this fourth-degree equation by a small positive number. In this way we perturb the curve (which is nonsmooth at four points) into a curve that becomes smooth near these points. As a result we obtain a smooth curve of degree 4 with four connected components:

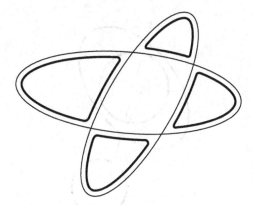

This curve solves our problem. Each of the four connected components bounds a disc in the plane that does not contain any of the other connected components inside it.

It is possible for a disc bounded by one of the components of a smooth curve of degree 4 to contain other components inside it, as shown, for example by the curve $(x^2 + y^2 - 1)(x^2 + y^2 - 2) = 0$.

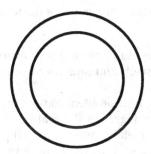

Problem. Can the above situation occur for a degree-4 curve consisting of four (or three) connected components?

Solution. The restriction of a degree-4 polynomial in two variables to a line has degree at most 4. Therefore, every line intersects such a curve in at most four points.

If the disc bounded by one of the components were to contain another component (and therefore the disc bounded by it), then the line joining a point A of this smaller disc to a point B of the third component would intersect the curve more than four times, which is impossible.

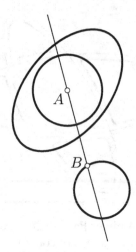

Therefore there cannot be a third component.

Thus, an algebraic curve of the fourth degree in the real projective plane consisting of four connected components is topologically standard: its four connected components bound four nonintersecting discs, as in the example constructed above; no other arrangement of the four connected components is possible.

For certain reasons, the study of curves of even degree is simpler than that of curves of odd degree,[12] and thus we consider curves of degree $n = 6$ as the next example.

In the statement of his 16th problem (on the topological structure of algebraic curves of a given degree in the real projective plane), Hilbert announced that he had completely investigated all possible arrangements of the connected components of a degree-6 curve under the assumption that the number of components reaches its maximum value of $M = 11$, given by Harnack's theorem.

According to Hilbert's statement there are only two such arrangements.

Hilbert did not publish a proof of his theorem, but a proof was published about 70 years later by Dmitry Andreevich Gudkov, a mathematician from Nizhny Novgorod. Gudkov was a student of the remarkable physicist A. A. Andronov. Working on the theory of oscillations and the theory of bifurcations, Andronov naturally came to a study of similar topological questions (and not just for polynomials).

Shortly thereafter, I. G. Petrovsky, who in the 1930s had obtained some remarkable results on Hilbert's 16th problem, asked me to referee Gudkov's habilitation thesis, in which he refuted both Hilbert's "theorem" and his own earlier proof of it.

Gudkov's striking result was that the number of possible arrangements was three, and not two as claimed by Hilbert.[13]

Here are the three arrangements of the 11 ovals of an algebraic sixth-degree curve in the real projective plane:

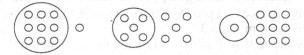

In other words, of the 11 ovals (each of which divides $\mathbb{R}P^2$ into a topological disc and a topological Möbius strip), only one oval contains other ovals inside the disc bounded by it; moreover, the number of these interior ovals can be only 9, 5, or 1 (and all these three cases are realized).

Hilbert had claimed that the case of five interior ovals was impossible, but he was wrong.

The total number of possible topologically distinct arrangements of 11 closed nonintersecting smooth curves runs to many billions.[14] Gudkov's striking result excluded these billions not only in the case of curves of degree 6. In his thesis he found hundreds of results on the impossibility of a number of arrangements of the ovals of curves of higher degree.

In considering these examples, I discovered that the number 8 often plays a special role in them. For example, in the Gudkov curves depicted above, the possible numbers of interior ovals (9, 5, and 1) are in arithmetical progression with common difference 4, while the Euler characteristics[‡] of the set F of points where $f(x,y) \leq 0$ (where we assume that the polynomial f of degree 6 defining the curve is positive for large x and y) are equal to $-7, +1$, and $+9$, respectively.

For a polynomial f of degree $2k$ that is positive at infinity and has $M = g + 1$ ovals, the general statement is this:

$$\text{(Euler characteristic of the set } F \text{ where } f \leq 0) \equiv k^2 \ (\text{mod } 8).$$

In my review of Gudkov's habilitation thesis I called this congruence, which was proved by him for many curves, "Gudkov's conjecture" (although he tried to convince me that the congruence does not always hold).[15]

On the other hand, I knew that congruences modulo 8 play a special role in the topology of smooth 4-dimensional manifolds and started looking for such a manifold in Hilbert's problem on algebraic (1-dimensional) curves.

After much effort, I discovered that the required manifold is the complexification of the surface F that is defined above by $f(x,y) \leq 0$ and whose boundary is the algebraic curve given by the equation $f(x,y) = 0$.

To complexify the condition $f(x,y) \leq 0$ defining the surface, I wrote it in the form $f(x,y) + z^2 = 0$. This equation defines (for complex x, y, and z) a

---

[‡] The Euler characteristic of a disc is 1, while for a disc with a holes it is $1 - a$. The Euler characteristic of several disconnected regions is the sum of the Euler characteristics of the components. The Euler characteristic of the printed letters of the word "YES" is three, while the characteristic of the word "NO" is one.

double cover of the complex projective plane $\mathbb{C}P^2$ ramified in the algebraic curve given by the equation $f(x,y) = 0$.

By applying results of the topology of 4-manifolds to this manifold (complexifying the surface F), I proved Gudkov's conjecture on congruence, but only modulo 4, not modulo 8.

Using my proof, V. A. Rokhlin completed the proof of Gudkov's conjecture modulo 8;[16] this achievement stands today as one of the central results related to Hilbert's 16th problem.

This advance in real algebraic geometry has thus grown from investigations of Hilbert's question on the topology of plane algebraic curves of degree 6.

For degree-8 curves these results also gave many restrictions, cutting down the number of possible arrangements of the 22 ovals to fewer than 100—at present, this number has been reduced to approximately 80 arrangements.

But only (approximately) 70 topologically different configurations have been realized by algebraic curves in the real projective plane. There remain about a dozen cases[17] for which none of the known restrictions (including Gudkov's congruence and the count of points of intersection with lines)[18] excludes the existence of an algebraic curve of degree 8 in the projective plane $\mathbb{R}P^2$ with such an arrangement of the 22 ovals, although no example of such a curve has been constructed.

This is one of those problems of mathematics that relate to its very foundations and are required for an enormous number of applications (the laws of nature are often described by equations of the form $f(x,y) = 0$, and the disposition of the branches of such manifolds provides qualitative inferences about the natural phenomena described by these laws).

The results described have linked this problem not only with the modern topology of smooth manifolds, but also with such areas of mathematics and physics as symplectic topology and contact geometry, quantum field theory, and (arising in it) new invariants of smooth manifolds, knots, and other objects.

The theory of invariants of knots was suggested by Kelvin for explaining Mendeleev's periodic table by the microscopic but stable topological differences in the fine structures of atoms. Today's topological methods of quantum field theory and elementary particle theory are a kind of development of this idea, and the emerging discrete and topological invariants of the different objects of mathematical physics and algebraic geometry (of which real algebraic curves in the projective plane provide only the simplest model) are leading to rapid advances in many areas of fundamental mathematics.

The question of Hilbert's 16th problem on algebraic curves is not the only natural problem of real algebraic geometry. For example, just as natural (and even more essential for applications of mathematics) is the question of the topological classification of functions f (for example, defining curves $f(x,y) = 0$).[§]

We consider two functions $f_0 : M \to \mathbb{R}$ and $f_1 : M \to \mathbb{R}$ to be topologically equivalent[19] if one can be converted into the other by a continuous deformation $\{f_t\}$, $0 \le t \le 1$, where all intermediate functions f_t are taken into each other by continuously varying smooth substitutions of the independent variables ($g_t : M \to M$) and the dependent variables ($h_t : \mathbb{R} \to \mathbb{R}$) so that

$$f_t(x) = h_t(f_0(g_t(x))) \text{ for all } t \text{ from } 0 \text{ to } 1.$$

Example. The polynomials $f_0(x) = x^4 - 2x^2 + Ax$ and $f_1(x) = x^4 - 2x^2 - Ax$ with $A > 0$ can be topologically inequivalent even though they both have three critical points (where $df/dx = 0$) and each remains equivalent to itself under a small change of the coefficient of the x^4 term.

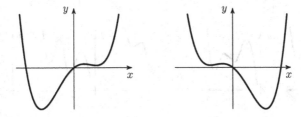

In the case $A = 0$, one obtains a function that is not topologically equivalent to either of the above two functions (this function also has three critical points) and that additionally, changes its topological type under certain small changes of the coefficients (since the coincidence of the values at two of the critical points, which occurs for $A = 0$, can be destroyed by an arbitrarily small perturbation $\tilde{f}$).

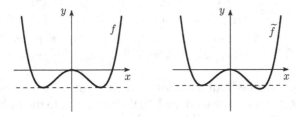

Problem. How many topologically different polynomials $f : \mathbb{R} \to \mathbb{R}$ of degree $n + 1$ with n critical points can exist? (A nonconstant degree-$(n + 1)$ polynomial cannot have more than n critical points.)

Hint. Draw the graphs of polynomials with leading term x^{n+1} with n distinct critical points and with n distinct critical values for small n:

[§] A survey of recent investigations along these lines is given in V. I. Arnol'd, "Topological classification of Morse functions and generalizations of Hilbert's 16th problem," *Math. Phys. Anal. Geom.* 10 (2007), no. 3, 227–236.

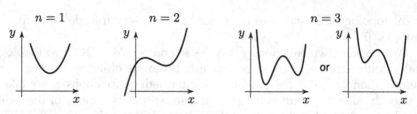

The corresponding numbers $K(n)$ of topologically inequivalent polynomials of this form are

$$K(1) = 1, \quad K(2) = 1, \quad K(3) = 2.$$

Continuing the drawing of graphs for higher-degree polynomials, we find, for example for $n = 4$, the following $K(4) = 5$ topologically inequivalent possibilities for degree-5 polynomials with leading term x^5 and with four critical points:

Continuing with such drawings, it is easy to work out that $K(5) = 16$, $K(6) = 61$, $K(7) = 272$. These numbers can be computed with the aid of the following *Euler–Bernoulli triangle*:

					1					
				0		1				
			1		1		0			
		0		1		2		2		
	5		5		4		2		0	
0		5		10		14		16		16
61	61		56		46		32		16		0
0	61	122		178		224		256		272	272

$K(1) = 1$
$K(2) = 1$
$K(3) = 2$
$K(4) = 5$
$K(5) = 16$
$K(6) = 61$
$K(7) = 272$

This triangle is filled out as follows: in each row of even rank (second, fourth, and so on) we write 0 on the left; then, moving to the right, we write in each position the sum of all the numbers to the left in the previous row (for example, for the fourth row we write $\{0, 1, 1 + 1 = 2, 1 + 1 + 0 = 2\}$).

The odd rows are filled out in similar fashion, except that we start with a zero on the extreme right; we then move leftwards, writing in each position the sum of all the numbers to the right in the previous row (for example, for the fifth row we write consecutively $\{0, 2, 2 + 2 = 4, 2 + 2 + 1 = 5, 2 + 2 + 1 + 0 = 5\}$).

It is not difficult to prove that the sum of the numbers placed in the nth row gives the number $K(n)$ of topologically inequivalent polynomials $x^{n+1} + \cdots$ with n distinct critical points and n distinct critical values.

On the other hand, the Euler–Bernoulli triangle gives rise to the following remarkable formula for $K(n)$.

Theorem. The sum of the power series[¶]

$$H(t) = \sum_{n=0}^{\infty} K(n)\frac{t^n}{n!}$$

is $H(t) = \sec t + \tan t$.[20]

Thus, the left-hand side of the Euler–Bernoulli triangle gives the numbers $1, 1, 5, 61, \ldots$, which determine the coefficients of the power series

$$\sec t = \frac{1}{\cos t} = 1 + 1\cdot\frac{t^2}{2!} + 5\frac{t^4}{4!} + 61\frac{t^6}{6!} + \cdots.$$

The numbers $1, 1, 5, 61, \ldots$ are called the *Euler numbers*. This sequence is easily recognized; the number 61 is rarely encountered in other sequences (while the beginning $1, 1, 5, \ldots$ is not so rare).

The right-hand side of the Euler–Bernoulli triangle gives the numbers $1, 2, 16, 272, \ldots$, which determine the coefficients of the power series

$$\tan t = t + 2\frac{t^3}{3!} + 16\frac{t^5}{5!} + \cdots.$$

By dividing the sine expansion by the cosine expansion, one easily sees that the power series for the tangent indeed begins with the terms

$$\tan t = t + \frac{t^3}{3} + \frac{2t^5}{15} + \cdots$$

(where $1/3 = 2/3! = 2/6$, $2/15 = 16/5! = 16/120$, and so on).

The proof of the above theorem is easiest to obtain by tearing the graph of a polynomial $x^{2m} + \cdots$ with $n = 2m - 1$ critical points (with distinct critical values) at the absolute minimum point.

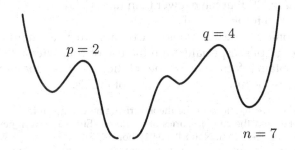

_____

[¶] Series of this form are called the *exponential generating functions* of the sequences $K(0), K(1), \ldots$, because for $K(n) \equiv 1$ one obtains the series for e^t.

Then by considering the topological types of both graphs thus obtained with p and q critical points (where $p + 1 = n$) and expressing the number of distinct topological types of the original function in terms of the number of topological types of both parts (but one must take into account the disposition of the critical values at the critical points of the left-hand part among the critical values at the critical points of the right-hand part), one can prove[||] a relation between the number $K(n)$ and the products $K(p)K(q)$, where $p + q = n - 1$. This relation is expressed in terms of the exponential generating function H in the form of a differential equation whose solution (with initial condition $H(0) = 1$) gives the formula for H:

$$2\frac{dH}{dt} = H^2 + 1.$$

For polynomials in several variables, the situation is more complicated. For example, for degree-4 polynomials in two real variables there are at most nine critical points. Such polynomials with nine critical points that behave at infinity like $x^4 + y^4$ can be arranged into 17,746 classes of topologically different types.

All these classes are realized by smooth functions with this number of critical points. But whereas for the case of functions of one variable, all the topologically inequivalent types are realized by polynomials of the required degree, this is not so for functions of two variables.

How many topologically different fourth-degree polynomials in two variables (behaving like $x^4 + y^4$ at infinity) are there with nine critical points in $\mathbb{R}^2$? I do not know the answer to this question, but I think that the answer is less than a thousand.

The number of topological types of smooth functions with n critical points in $\mathbb{R}^2$ increases with n at the same rate as n^{4n}. But how the number of classes realized by polynomials increases is unknown—one cannot rule out the possibility that it is only of polynomial growth[**] of order n^{const}.

All these questions as well as the closely related question on the number of regions into which the space of polynomials in m variables of degree d is divided by the manifold of degeneracies (formed by the set of all polynomials of degree d in $\mathbb{R}^m$ that have fewer than $(d-1)^m$ different critical values) remain open even for $m = 2$ and $d = 4$.

I am mentioning these questions here because they are in principle accessible to school pupils, especially with the use of computers; thus I hope in the future to obtain information on the solutions of these questions from the audience of these lectures of mine for school pupils.

[||] A survey of the investigations of the theory arising here is given in V. I. Arnol'd, "The calculus of snakes and the combinatorics of Bernoulli, Euler and Springer numbers of Coxeter groups," Uspekhi Mat. Nauk 47 (1992), no. 1, 1–51.

[**] In this connection, see the papers V. I. Arnol'd, "Smooth Functions Statistics," *Funct. Anal. Other Math.* 1 (2006), no. 2, 111–118, and L. Nicolaescu, "Morse Functions Statistics," *Funct. Anal. Other Math.* 1 (2006), no. 1, 85–91.[21]

Here is yet another such question on which progress was made thanks to computer calculations.

Consider the graph of the function $\{z = f(x,y)\}$ as a surface in Euclidean 3-space $\mathbb{R}^3$.

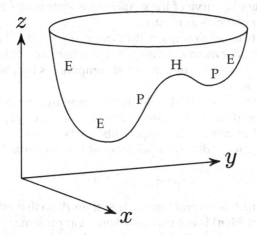

In the neighbourhoods of some points, this surface is locally convex. We call such points *elliptic*; an example is the surface of an ellipsoid.

In the neighbourhood of other points, the surface intersects its tangent planes in a pair of intersecting lines (and is locally nonconvex). Such points are called *hyperbolic* (an example is the surface of a hyperboloid of one sheet—all its points are hyperbolic).

The regions of elliptic and hyperbolic points of a surface are separated by a *curve of parabolic points*.

The equation of this curve has the form $H(x,y) = 0$, where the function H (called the *Hessian* of the function f) is the determinant formed by the second partial derivatives of f:

$$H = \begin{vmatrix} \dfrac{\partial^2 f}{\partial x^2} & \dfrac{\partial^2 f}{\partial x \partial y} \\ \dfrac{\partial^2 f}{\partial y \partial x} & \dfrac{\partial^2 f}{\partial y^2} \end{vmatrix} = f_{xx}f_{yy} - f_{xy}^2.$$

If f is a polynomial of degree n, then H is a polynomial of degree $2n - 4$. (For $n = 4$, the degree of the Hessian is 4, so that the parabolic curve is given in the (x,y)-plane by an algebraic equation of degree 4.)

Problem. How many connected components can a parabolic curve have if f is a polynomial of degree n (for example, when $n = 4$)?

Naturally, a parabolic curve can be completed by points at infinity, by considering the plane $\mathbb{R}^2$ with coordinates x,y to be the affine part of the projective plane $\mathbb{R}P^2$. Thus there are two questions concerning the number of components: one can consider the components in the affine plane (where for the hyperbola there are two), or one can consider the components in the projective plane (where the hyperbola is connected).

It follows from Harnack's theorem that the number of connected components of the parabolic curve of the graph of a polynomial f of degree $n = 4$ (in the projective plane) is at most 4.

One can construct an example with three components. (I leave this construction to the reader as an exercise.) On the other hand, it is hard to determine whether the case of four connected components is achieved for some polynomial f of degree 4.

My Mexican student A. Ortiz-Rodriguez recently proved in her dissertation, which she defended at the University of Paris VI, that the maximum number M of connected components (in the projective plane $\mathbb{R}P^2$) of a parabolic curve for a degree-n polynomial in two variables satisfies the inequalities

$$An^2 \leq M(n) \leq Bn^2.$$

Here the constant B is several times larger than A, so that even the asymptotic behaviour of $M(n)$ is not precisely known at present.[22]

For the case $n = 4$, Ortiz-Rodriguez devoted a year's continuous work on a computer in Mexico investigating fifty million $(5 \cdot 10^7)$ different degree-4 polynomials.

Among them there proved to be three polynomials that had in their graph a parabolic curve with four connected components.

When the computer gave out the values of the coefficients of these polynomials, the verification of the number of their connected components took a few minutes, even without any further help from the computer.

Thus Ortiz-Rodriguez obtained a rigorous mathematical proof of the relation $M(4) = 4$ "formally without using a computer."

Yet finding this proof without a computer had long been unsuccessful (furthermore, four connected components were observed in only three of the fifty million cases).

Here I have mentioned the use of a computer (so far, the only one) in real algebraic geometry (the theory of parabolic curves is related to symplectic geometry, too) because I hope for progress from my listeners and readers in the investigation of the behaviour of the number of components $M(n)$ as $n \to \infty$.

For smooth algebraic surfaces of degree n in projective space $\mathbb{R}P^3$ (given in Cartesian coordinates of the affine part of the projective space by equations of the form $f(x,y,z) = 0$, where f is a polynomial of degree n), upper and lower estimates of the number $P(n)$ of connected components of parabolic curves in Ortiz-Rodriguez's dissertation have the form

$$Cn^3 \leq P(n) \leq Dn^3$$

(surfaces of degree n with at least Cn^3 closed connected parabolic curves exist; surfaces of degree n with more than Dn^3 closed connected parabolic curves do not exist).[23]

In the above theorem, the ratio D/C is of order 20. The question of the true asymptotic behaviour of the maximum number of connected components of a parabolic curve of a smooth algebraic surface of degree n in $\mathbb{R}P^3$ is open: probably the number increases as En^3, but there is no precise conjecture concerning the constant E; nor has there been any empirical computer investigation based on calculations of $P(n)$, say for $n \leq 100$. This question I also leave to my listeners and readers.

Chapter 5
Complex Algebraic Curves

The equation $f(x,y) = 0$, where x and y are complex variables and f is a polynomial (with complex coefficients in general), defines a subset of the complex plane $\mathbb{C}^2$ (with coordinates x and y) of dimension two, since the real dimension of a 2-dimensional complex plane is equal to 4, and equating to zero the complex number $f(x,y)$ means equating to zero both its real part and its imaginary part; that is, we have two equations in the four real variables $(\operatorname{Re} x, \operatorname{Im} x, \operatorname{Re} y, \operatorname{Im} y)$.

Example. The "complex circle" is given in $\mathbb{C}^2$ by the equation $x^2 + y^2 = 1$.

Problem. Investigate the topological structure of this two-dimensional (in the real sense) subset of the four-dimensional (in the real sense) space $\mathbb{C}^2$.

Solution. Multiplying the y coordinate by i, we get the equation of the circle in the form of the equation for the hyperbola $x^2 - z^2 = 1$. Thus, from the point of view of complex geometry, the circle and the hyperbola are one and the same curve; what is different is merely the coordinates in the complex plane in which the equation of this curve is written. By the further change of coordinates $(X = x + z, Y = x - z)$, we arrive at the equation of a hyperbola in another standard form, namely $XY = 1$, or $Y = 1/X$.

It is clear from this equation that the complex circle is topologically equivalent to the plane of the complex variable X without the origin, that is, to the cylinder $S^1 \times \mathbb{R}$.

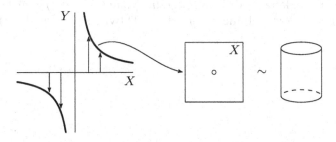

V.I. Arnold, *Real Algebraic Geometry*, UNITEXT – La Matematica per il 3+2 66,
DOI 10.1007/978-3-642-36243-9_5, © Springer-Verlag Berlin Heidelberg 2013

As X approaches zero, Y tends to infinity, and as X tends to infinity, Y approaches zero.

It is easy to deduce from this that in the complex projective plane $\mathbb{C}P^2$, the complex circle or hyperbola (just as is the real hyperbola in $\mathbb{R}P^2$) is completed by two points at infinity (one of which goes off in the direction of the X-axis in the (X,Y) plane, and the other in the direction of the Y-axis).

It is clear from this that the complex circle (or hyperbola) is topologically equivalent (homeomorphic and even real diffeomorphic) to the two-dimensional real sphere:

Just as is the case on the real hyperbola in $\mathbb{R}P^2$, the intersection points with the line of points at infinity (that is, the asymptotic directions of the hyperbola in the affine plane) are not any worse than ordinary points; the complex circle (or hyperbola) is as good in neighbourhoods of each of its two points at infinity as in neighbourhoods of any point in the (X,Y)-plane: from the real point of view, it is a smooth two-dimensional surface in real four-dimensional space.

Problem. Prove that all real circles extended to curves in $\mathbb{C}P^2$ by adding complex points and points at infinity have two points in common.

Solution. We calculate the points at infinity on the complexification of the circle given by the equation $x^2 + y^2 = 1$. These are the asymptotic directions $(z = \pm x)$ of the hyperbola $x^2 - z^2 = 1$, that is, those points at infinity in the complex projective plane $\mathbb{C}P^2$ that the points of the complex plane $\mathbb{C}^2$ with coordinates $(x = t, y = it)$ and $(x = t, y = -it)$ approach as $t \to \infty$.

For any other circle (always, in fact, with an equation of the form $(x - a)^2 + (y - b)^2 = r^2$, the asymptotic directions are the same. Therefore the complex points at infinity of all circles are the same.

Another description of the complex circle is provided by a rational parameterization of it.

In the (x,y)-plane we draw through the point $(x = -1, y = 0)$, which lies on the circle $x^2 + y^2 = 1$, the line $y = t(x + 1)$ (with slope equal to t).

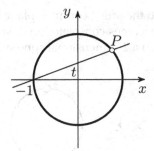

The second point of intersection P of this line with the circle can be obtained from the equation $x^2 + t^2(x+1)^2 = 1$, that is,

$$(1+t^2)x^2 + 2t^2x + (t^2 - 1) = 0.$$

One of the roots of this quadratic equation is already known, namely, $x = -1$ (since the line passes through the point $(-1,0)$). By Viète's theorem, the other root is given by

$$x = 1 - \frac{2t^2}{1+t^2} = \frac{1-t^2}{1+t^2}.$$

Hence it follows that

$$y = t(x+1) = \frac{2t}{1+t^2}.$$

The parameterization so obtained,

$$x = p(t) \quad y = q(t), \quad \text{where } p = \frac{1-t^2}{1+t^2}, \ q = \frac{2t}{1+t^2},$$

establishes a one-to-one correspondence between the complex projective circle (completed by its two points at infinity) and the complex projective line with axis t (completed by one point at infinity).

This correspondence establishes not only a topological equivalence (homeomorphism) between the complex projective circle and the complex projective line (which, as we shall soon see, is a 2-dimensional sphere from the real point of view).[24] This homeomorphism is, in fact, a diffeomorphism (that is, a map that together with its inverse is smooth), and this is true from both the real and complex points of view.

Problem. Investigate the topological properties of the complex line completed by its points at infinity that appear when one extends the $\mathbb{C}^2$ plane to the complex projective plane $\mathbb{C}P^2$.

Solution. In the complex plane $\mathbb{C}^2$, the affine complex line is in the real sense a 2-dimensional plane in real 4-dimensional space (as is clear, for example, for the complex x-axis of the plane $\mathbb{C}^2$ with complex coordinates x

and y). When one adds to the affine complex plane $\mathbb{C}^2$ the points at infinity, the affine complex line becomes completed by just one point. Such a completion converts the line to a 2-dimensional sphere:

$$\mathbb{C} + \{\infty\} = \bigcirc \sim S^2$$
$$\mathbb{C}P^1$$

Algebraic curves homeomorphic to the sphere ("having zero genus") are called "rational curves," since they can be defined parametrically by equations of the form $\{x = p(t), y = q(t)\}$, where p and q are rational functions. For example, for the circle, a suitable parameterization is

$$x = \frac{u^2 - v^2}{u^2 + v^2}, \quad y = \frac{2uv}{u^2 + v^2}$$

(where $t = \frac{v}{u}$), as derived above. For integral values of the variables (u, v) it gives rational points of the circle; for example, for $u = 1$, $v = 2$ we get the "Egyptian" triangle $(x = 3/5, y = 4/5)$.

Remark. The point at infinity on the complex projective line in no way differs from its other points. In a neighbourhood of the point at infinity "$x = \infty$" one can choose as the local coordinate on $\mathbb{C}P^1$ the coordinate $y = 1/x$ (in a neighbourhood of the point $y = 0$).

Problem. Consider the subdivision of the affine complex x-axis into real "horizontal" lines along which the imaginary part of x is constant:

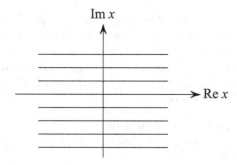

Draw this family of real lines in a neighbourhood of the point $y = 0$, where y, the local coordinate in a neighbourhood of this point, is equal to $1/x$ outside this point (and outside the point $x = 0$).

Solution. The transformation $x \mapsto 1/x$ replaces the modulus of a complex number by the inverse of that real number, and changes the sign of the argument of x. Therefore this transformation is the composition of the inversion in the circle of radius 1 with centre at the point $x = 0$ and a reflection of the plane of the y variable in the line $\mathrm{Im}\, y = 0$.

Under the inversion, lines parallel to the horizontal line $\mathrm{Im}\, x = 0$ are taken into circles passing through the point $y = 0$ and tangent (at this point) to the line $\mathrm{Im}\, y = 0$. Therefore, the entire family of horizontal lines in the plane of the complex variable x turns, in the plane of the complex variable $y = 1/x$, into a family of circles tangent to each other at the origin:

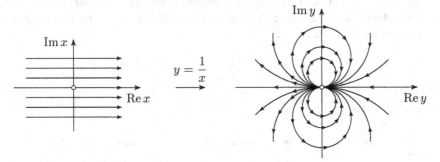

As in the previous examples, algebraic curves defined in the complex affine plane $\mathbb{C}^2$ with complex coordinates (x,y) by algebraic equations of the form $f(x,y) = 0$, where f is a polynomial (with complex coefficients in general), are, from the real point of view, two-dimensional surfaces in real four-dimensional space going off to infinity in certain directions.

Under completion of the complex affine plane $\mathbb{C}^2$ by (complex) points at infinity to the complex projective plane $\mathbb{C}P^2$, these two-dimensional surfaces become completed by their own points at infinity, so that one obtains closed two-dimensional surfaces. These surfaces, like the completed cubic parabola in the real case, may have singularities.

However, in what follows we shall consider mainly the case in which there are no singularities. In this case the surface described is closed and connected. It is called a Riemann surface (since such surfaces with their non-trivial topology had already been considered by Newton, in Lemma 28 of his *Mathematical Elements of Natural Philosophy*).[25]

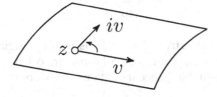

We note first of all that all these smooth real two-dimensional surfaces are orientable. More precisely, if a (nonzero) vector v is tangent to this surface at a point z, then the vector iv is also tangent to this surface at the same point; therefore the pair of vectors (v, iv) defines an orientation of the surface (at the point z and in a sufficiently small neighbourhood of it).

These orientations (in the neighbourhoods of different points) are compatible, because the operation of multiplying tangent vectors by i depends continuously on the point z of tangency. (This is true not only for points of the affine complex plane $\mathbb{C}^2$, but also for the points at infinity that get appended in passing to the complex projective plane $\mathbb{C}P^2$, since multiplication of tangent vectors by the complex number i is defined at those points as well.)

In topology, all two-dimensional connected closed orientable surfaces are classified as follows. Each of them is diffeomorphic to one (and exactly one) surface in the following list:

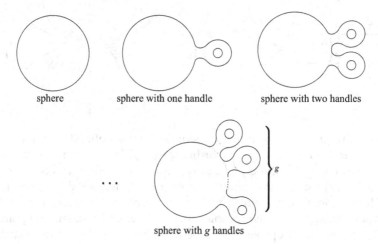

sphere   sphere with one handle   sphere with two handles

sphere with g handles

Example. The surface of a torus (car tyre) is topologically equivalent to a sphere with one handle:

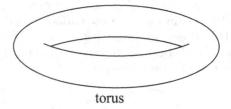

torus

Example. The real projective plane $\mathbb{R}P^2$ is not topologically equivalent to any of the surfaces in the above list: all those are orientable, but the projective plane is nonorientable (since it contains the Möbius strip).

Problem. Is the n-dimensional real projective plane

$$\mathbb{R}P^n = \frac{\mathbb{R}^{n+1} \setminus 0}{\mathbb{R} \setminus 0} = \frac{S^n}{\{+1,-1\}},$$

obtained from the n-sphere S^n by identifying diametrically opposite points, orientable?

Answer. The spaces $\mathbb{R}P^1$, $\mathbb{R}P^3$, $\mathbb{R}P^{2k-1}$ (of odd dimension) are orientable, while the spaces $\mathbb{R}P^2$, $\mathbb{R}P^4$, $\mathbb{R}P^{2k}$ (of even dimension) are nonorientable.

For a sphere with g handles, the number g is called the *genus* of the surface. For example, the genus of a sphere is equal to zero, the genus of a torus is equal to one, and the genus of a pretzel is equal to two.

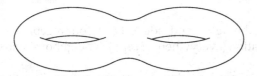

Theorem (Riemann–Hurwitz). *A smooth algebraic plane curve of degree n in the complex projective plane $\mathbb{C}P^2$ has a Riemann surface of genus*

$$g = \frac{(n-1)(n-2)}{2}. \tag{3}$$

Remark. We have already seen above that complex projective curves of degrees 1 and 2 (lines and circles) are spheres from the topological point of view (they have genus zero).

This makes it easy to remember formula (3), especially if one also uses the simple geometric solution of the following problem.

Problem. Find the genus of the Riemann surface of a smooth algebraic curve of degree 3 in the complex projective plane defined in affine coordinates (x,y) in the affine plane $\mathbb{C}^2$ by the equation $xy(1-(x+y)) = 1/100$.

Solution. If instead of $1/100$ we had zero on the right-hand side of the above equation, then the equation would define three sides of a triangle.

Each of the sides is (in the complex projective plane) a complex projective line, which is topologically a 2-sphere.

But the whole complex triangle is not a smooth two-dimensional surface, since the three spheres (a,b,c) forming its sides pairwise intersect (at the vertices A, B, C):

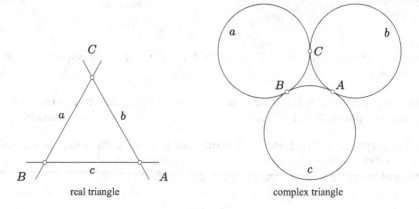

real triangle   complex triangle

On changing the right-hand side to $1/100$, our algebraic curve undergoes a small perturbation (everywhere except the neighbourhoods of the singular points A, B, C).

At each of the singular points, the singular curve (two branches of which intersect at this point) undergoes the same modification that happens with a pair of intersecting lines given by the equation $xy = 0$ when it is replaced by the hyperbola $xy = 1/100$.

This modification consists in the following. At each of the two intersecting branches, a small neighbourhood of the intersection point is removed, and then both these neighbourhoods are replaced by a cylinder joining the boundaries of the removed neighbourhoods (as we saw for the hyperbola on p. 55:

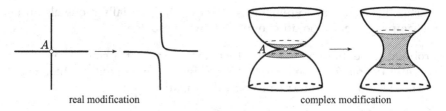

real modification   complex modification

Having made this modification at each of the three vertices of our triangle ABC, we obtain from it the following smooth surface (in which the neighbourhoods of the three vertices A, B, C are replaced by small cylinders that join the spheres intersecting at these vertices):

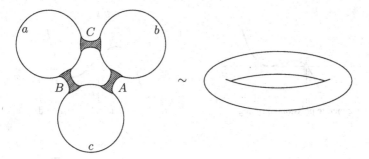

As is clear, the above surface has genus $g = 1$; topologically, it is equivalent to a torus (or a sphere with one handle). Algebraic curves of genus 1 are called elliptic curves (since they are related to "elliptic integrals," line integrals expressing the arc length of an ellipse).

Knowing that $g(3) = 1$, we recover the denominator 2 in the Riemann–Hurwitz formula (3): in the case of degree $n = 3$, the product $(n - 1)(n - 2)$ is equal to 2, so that one needs to divide it by 2 to obtain the genus $g = 1$ for our smoothed triangle.

The proof of formula (3) in the general case follows the same method, but I preface it with a remark of a general character.

Mathematics, as in general all sciences, is international; there exists no specific German mathematics, such as the Nazis aspired to, nor a specific French or Russian mathematics (although many people think that every problem admits two extreme versions: the Russian version that nobody can simplify without making it trivial, and the French version that nobody can generalize any further).

But in algebraic geometry there is a general principle that emerged a long time ago and is traditionally called the Italian principle (because it was systematically used by the remarkable Italian algebraic geometers of the nineteenth century).

Here then is the "Italian principle." Suppose that we are considering some family of complex objects (for example, the manifold $\mathbb{C}^n$ of all complex polynomials of degree n of the form

$$x^n + a_1 x^{n-1} + \cdots + a_n$$

or the manifold of all plane algebraic curves of degree n in the complex projective plane $\mathbb{C}P^2$).

Some objects of the family will then be nondegenerate (for example, polynomials without multiple roots or algebraic curves without singular points). Other objects are degenerate:

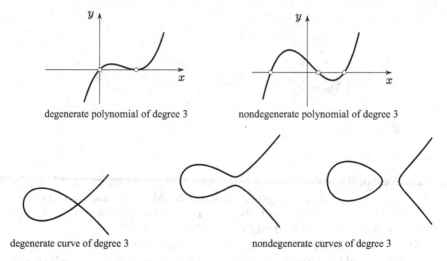

degenerate polynomial of degree 3   nondegenerate polynomial of degree 3

degenerate curve of degree 3   nondegenerate curves of degree 3

The "Italian principle" consists in the following: *all nondegenerate objects in a family of complex objects have the same topological structure* (for example, all nondegenerate polynomials of degree n have the same number of roots and all smooth algebraic curves of degree n in the complex projective plane have the same genus g).

Strictly speaking, to make this principle into a theorem we should define rigorously what the words "family," "nondegenerate," and "same topological structure" mean. This could be done; but that would require many pages of writing. Therefore I prefer to explain the reason for the above phenomenon, leaving the details to inquisitive readers.

The main observation is that degeneracy is expressed by the vanishing of some complex polynomial in the parameters of the objects of the family under consideration.

For example, a polynomial has multiple roots if and only if its discriminant (polynomially expressed in terms of the coefficients of that polynomial) vanishes. In precisely the same way, an algebraic curve defined in the projective plane by an equation $f(x,y) = 0$, where f is a polynomial of degree n, has singularities if and only if some polynomial in the coefficients of f vanishes.

The vanishing of a complex polynomial in the parameters of the objects under study means the vanishing of both the real and the imaginary parts of that polynomial. Therefore, one such complex condition of degeneracy singles out in the space of all objects under study a set of degenerate objects having codimension 2, rather than 1 (as is the case for points in the plane or curves in space).

The set of degenerate objects having (real) codimension 2 does not decompose the entire space of complex objects under study: any two nondegenerate points A and B of this space can be joined by a continuous path γ

not intersecting the set of degenerate objects (and avoiding this set of codimension two).

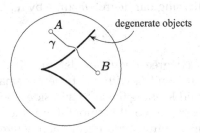

In moving along such a nondegenerate path, the topological characteristics considered (the number of roots of a polynomial, the genus of a curve, and so on) do not change.

As a consequence of this, for the calculation of these common characteristics of the objects of the family it suffices to study one example of a nondegenerate object: its characteristics will provide information about all nondegenerate objects of the family.

Example. The degree-n polynomial

$$(x-1)(x-2)\cdots(x-n)$$

has n distinct roots and is nondegenerate.

Consequently, every degree-n polynomial with complex coefficients without multiple roots has exactly n complex roots.

We have thus proved the "fundamental theorem of algebra" (the property we proved for polynomials without multiple roots implies that polynomials of degree n with multiple roots also have at most n complex roots, since otherwise, together with an equation with more than n roots, some of which are multiple, one could produce an equation with more than n nonmultiple roots).

Applying the "Italian principle" to the proof of the Riemann–Hurwitz formula (3), we shall construct an example of a smooth degree-n algebraic curve in $\mathbb{C}P^2$ for which the genus can be explicitly calculated: $g = (n-1)(n-2)/2$.

In accordance with the "principle," all other nondegenerate (without singularities) degree-n algebraic curves have the same genus in $\mathbb{C}P^2$; and this proves formula (3).

To construct the example, we start with a set of n distinct lines in $\mathbb{C}P^2$. We choose them so that no three of them pass through one point. Then the total number of points of intersection is

$$N = (n-1) + (n-2) + \cdots + 1 = \frac{n(n-1)}{2}.$$

The union of our n lines is a degree-n algebraic curve with N singular points (at each of which two nontangent branches intersect).

We now smooth this singular degree-n curve by replacing its equation

$$f_1 f_2 \cdots f_n = 0 \tag{4}$$

(where $f_k(x,y) = 0$ is a degree-1 equation defining the kth of our n lines) by a similar degree-n equation but with a sufficiently small but nonzero ϵ on the right-hand side (and keeping the left-hand side of equation (4) intact).

The topological description of this smoothing is the same as in the above example of smoothing a complex triangle. It merely remains to calculate the genus of the surface that one obtains from the union of n two-dimensional spheres pairwise intersecting at N points when the neighbourhood of each point of intersection is replaced by a small thin tube joining the spheres.

To calculate this genus we start by smoothing the intersections of the first sphere with all the others.

After each such smoothing, one of the other spheres is combined with the first sphere into a single large sphere, so that finally, a single large sphere is obtained, which, however, still intersects itself at all points where the "other" spheres intersected each other.

The number of such intersection points (not touching the first sphere) is

$$N' = (n-2) + (n-3) + \cdots + 1 = \frac{(n-2)(n-1)}{2}.$$

Replacing each of these intersection points by a tube joining the intersecting branches, we glue N' handles to the original large sphere. This means that the result of all the smoothings is a surface of genus $g = N'$, which proves the Riemann–Hurwitz formula (3) for our smooth algebraic degree-n curve in the complex projective plane $\mathbb{C}P^2$ (and therefore, in accordance with the "Italian principle," also for all smooth algebraic degree-n curves in the complex projective plane $\mathbb{C}P^2$).

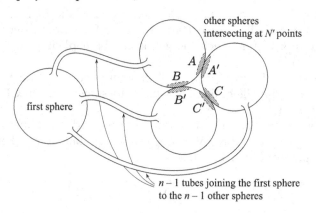

other spheres
intersecting at N' points

A
B A'
        C
B'
    C'

first sphere

$n-1$ tubes joining the first sphere
to the $n-1$ other spheres

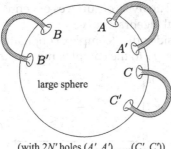

(with $2N'$ holes (A', A'), ...,(C', C'))

Thus, we have proved the Riemann–Hurwitz formula (3): the genus of a smooth algebraic degree-n curve in the complex projective plane $\mathbb{C}P^2$ is equal to

$$g = N' = \frac{(n-1)(n-2)}{2}.$$

The genus of the Riemann surface of a smooth algebraic curve in $\mathbb{C}P^2$ obtained by adding complex points to the real algebraic curve in $\mathbb{R}P^2$ gives an estimate of the number of connected components of the real algebraic curve by Harnack's theorem:

$$\text{number of components } \leq g + 1.$$

The proof of this theorem is based on the following topological property of surfaces of genus g: on such a surface there exist g (pairwise nonintersecting) closed curves that all together do not divide the surface (in other words, these curves have a connected complement), and this is the maximal number of such curves.

For example, the sphere is decomposed by just one closed curve, but the torus is not decomposed by a meridian, although any two nonintersecting closed curves on the torus certainly do divide it:

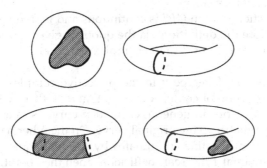

Furthermore, a surface of genus g is divided by any collection of $g + k$ nonintersecting closed curves into at least $k + 1$ parts. (A sphere is divided by k curves into $k + 1$ parts, while a torus can be divided by k curves into k parts.)

From this topological analysis of surfaces of genus g we derive Harnack's theorem concerning real algebraic curves.

To this end, we consider the "complex conjugation" map σ taking each complex number $z = x + iy$ to the complex number $\sigma(z) = x - iy$:

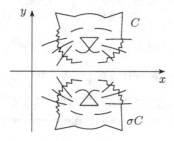

Each figure C in the plane $\mathbb{C}$ of the complex variable z is reflected in the real x-axis by σ, where $\sigma^2 = 1$, that is, $\sigma(\sigma z) = z$, $\sigma(\sigma C) = C$.

In just the same way, the *conjugation* $\sigma : \mathbb{C}^n \to \mathbb{C}^n$ (reflection of complex n-dimensional space in the real n-dimensional subspace) is defined by the same formula

$$\sigma(x + iy) = x - iy \quad \text{(for every } x \in \mathbb{R}^n, \, y \in \mathbb{R}^n).$$

In complex projective space, complex conjugation $\sigma : \mathbb{C}P^n \to \mathbb{C}P^n$ acts in the same manner. For example, one can extend it by continuity from the affine subspace to the "hyperplane at infinity" $\mathbb{C}P^{n-1}$. One can also act on the complex lines of the affine space $\mathbb{C}^{n+1}$ that pass through 0:

$$\mathbb{C}P^n = \frac{\mathbb{C}^{n+1} \setminus 0}{\mathbb{C} \setminus 0},$$

where on the right side the action of σ on the numerator determines its action on the quotient space.

The above action of σ on $\mathbb{C}P^n$ is continuous and is standard on its affine part $\mathbb{C}^n$. Therefore the definition via the quotient gives the same extension of the operation $\sigma : \mathbb{C}^n \to \mathbb{C}^n$ to the complex points at infinity as extension by continuity.

Algebraic curves of degree n in the projective complex plane $\mathbb{C}P^2$ are taken under the action of complex conjugation $\sigma : \mathbb{C}P^2 \to \mathbb{C}P^2$ to algebraic curves of degree n, but in general, each such curve is taken into another curve that is symmetric to the original curve. On the other hand, if the original curve is given in affine coordinates by the real equation $f(x,y) = 0$, where the polynomial f has real coefficients, then the operation of complex conjugation takes such a curve into itself.

In fact, $\sigma(uv) = \sigma(u)\sigma(v)$, and therefore

$$\sigma(f(x,y)) = (\sigma(f))(\sigma(x),\sigma(y)).$$

If the coefficients of the polynomial f are real, then $\sigma(f) = f$, while if f vanishes at the point (x,y), then

$$f(\sigma(x),\sigma(y)) = (\sigma(f))(\sigma(x),\sigma(y)) = \sigma(f(x,y)) = \sigma(0) = 0,$$

that is, the point $(\sigma(x),\sigma(y))$ reflected by the conjugation lies on the original algebraic curve.

Points of a real algebraic curve lying in the real projective plane $\mathbb{R}P^2$ stay unchanged under the conjugation symmetry $\sigma : \mathbb{C}P^2 \to \mathbb{C}P^2$. Other points of the curve move.

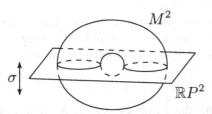

Therefore the connected components of a real algebraic curve in the real projective plane are the connected components of the set of points fixed by the complex conjugation symmetry σ on the Riemann surface M^2 of our curve.

These real closed curves divide the surface M^2 of genus g into parts that are surfaces with boundaries formed by the connected components of the real curve.

The symmetry σ takes each such part to another part of the same subdivision and with the same boundary.

If the number of parts is greater than two, then in addition to the two parts G and σG there are two further parts H and σH. The unions of G and σG and of H and σH are closed nonintersecting surfaces: $M = (G \cup (\sigma G)) \sqcup (H \cup (\sigma H))$.

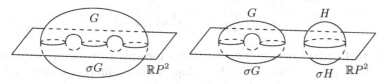

But a real smooth algebraic curve in the complex projective plane $\mathbb{C}P^2$ has a connected Riemann surface (this follows, for example, from the "Italian principle," since we proved this for the example with the perturbation of a union of lines).

Consequently, the number of parts into which the curves fixed by the conjuagation symmetry σ divide the Riemann surface M^2 of genus g is at most 2.

It follows from the topological properties of surfaces of genus g (see p. 68) that the number of connected components of a set of fixed curves cannot be large: if the number of components is equal to $g + k$, then the number of parts is at least $k + 1$. Therefore, the number of parts is at most 2 only for $k \leq 1$, that is, when the number of components $g + k$ is at most $g + 1$.

Hence, *the number of real components of a smooth curve of genus g in the real projective plane* $\mathbb{R}P^2$ *is at most* $g + 1$.

Remark. Harnack's theorem extends to real algebraic manifolds of larger dimensions in the complex projective space $\mathbb{C}P^n$, only instead of the genus g of a complex curve and the number of connected components of a real curve, one has to consider their multidimensional analogues called the "Betti numbers of nonoriented manifolds."

Without defining these topological invariants arising from homology theory, which for an m-dimensional manifold are nonnegative integers $b_0, b_1, \ldots, b_m$, I merely point out that for a 2-dimensional surface ($m = 2$) of genus g we have

$$b_0 = 1, \quad b_1 = 2g, \quad b_2 = 1$$

(the number b_k counts the number of independent k-dimensional cycles).

For a circle, the dimension m is 1, and the Betti numbers of a single circle are

$$b_0 = 1, \quad b_1 = 1.$$

For a union of r (pairwise nonintersecting) circles, the Betti numbers are

$$b_0 = r, \quad b_1 = r.$$

Generalizing Harnack's inequality, the multidimensional Smith inequality states that

$$\sum_{k=0}^{m} b_k(\mathbb{R}M^m) \leq \sum_{l=0}^{2m} b_l(\mathbb{C}M^m)$$

for any smooth m-dimensional algebraic manifold given by equations with real coefficients in projective space

$$\mathbb{R}M^m \subset \mathbb{R}P^n, \quad \mathbb{C}M^m \subset \mathbb{C}P^n.$$

The Betti numbers b_k and b_l in Smith's formula given above denote the "Betti numbers" of nonoriented homology theory with coefficients in $\mathbb{Z}_2$.

Example. For plane curves with Riemann surface of genus g ($m = 1, n = 2$), Smith's inequality takes the form

$$r + r \leq 1 + 2g + 1,$$

that is, the number r of connected components of a real curve is estimated as

$$r \leq g + 1.$$

In this case, Smith's inequality turns into Harnack's inequality,

The genus of algebraic curves considered above has many other applications apart from the above applications to the analysis of the number of components of a real projective curve.

Here is one such example. Let M be a rational algebraic curve in the plane with coordinates x and y, and R a rational function of x and y, that is, the ratio of two polynomials,

$$R(x,y) = \frac{a(x,y)}{b(x,y)}.$$

Then the "abelian integral"

$$I(X) = \int_{X_0}^{X} R(x,y)dx$$

along the rational curve M can be explicitly calculated in the form af an elementary function of the upper limit X (as a combination of rational functions, radicals, exponentials, trigonometric functions, and inverse trigonometric functions).

To prove this, it suffices simply to pass to the independent variable t in the integral (instead of x).

All second-degree curves are rational, and therefore the integrals of all rational expressions including $\sqrt{x^2 + ux + v}$ can be calculated in terms of elementary functions (for example, $\int dx/\sqrt{1-x^2} = \arcsin x$).

Example. For Newton's equation

$$\frac{d^2x}{dt^2} = F(x)$$

in a field with force $F(x)$ at the point x, we have the "law of conservation of energy"

$$\frac{d}{dt}\left(\frac{y^2}{2} + U(c)\right) = 0,$$

where $y = \frac{dx}{dt}$ and the potential energy U has at each point x a derivative equal to minus the force at this point:

$$\frac{dU}{dx} = -F.$$

If the field depends linearly on the coordinate x (as in Hooke's law, for example), then its potential energy is a second-degree polynomial. In the phase plane $\{(x,y)\}$, the motion in our field proceeds along a second-degree algebraic curve M:

$$\frac{y^2}{2} + U(x) = \text{const.}$$

The time of motion along this algebraic curve is expressed by an abelian integral:

$$y = \frac{dx}{dt}; \quad \text{therefore} \quad dt = \frac{dx}{dy};$$

$$t = \int_M \frac{dx}{y} = \int_{X_0}^X \frac{dx}{\sqrt{2(E - U(x))}}, \tag{5}$$

where E is the total energy (a constant of the law of conservation of energy).

Since second-degree curves are rational, this abelian integral can be expressed in terms of elementary functions if the field strength is linearly dependent on the position of the moving point.

For example, in the case of Hooke's law, calculating this abelian integral in terms of arcsines, we arrive at the harmonic oscillations

$$x = C_0 \sin(\omega t + \phi_0),$$

with constants C_0 and ϕ_0 depending on the initial conditions, and frequency ω depending on the coefficient in Hooke's law.

If, on the other hand, the force field is given by a polynomial in x of degree greater than 1, then the abelian integral (5) has to be calculated along an algebraic curve M,

$$\frac{y^2}{2} + U(x) = E,$$

of higher genus g (equal to 1 for polynomials U of degree 3 or 4, 2 for polynomials U of degree 5 or 6, and so on).

For example, for a cubic polynomial, when

$$2(E - U) = x^3 + g_2 x + g_3,$$

the abelian integral (5) along the elliptic curve M of genus 1 takes the form

$$t = \int_{X_0}^X \frac{dx}{\sqrt{x^3 + g_2 x + g_3}}$$

(with constant coefficients g_2 and g_3), called "Weierstrass normal form for elliptic integrals."

The function $X(t)$ inverse to $t(X)$ is called an *elliptic function*. It provides a remarkable generalization of ordinary trigonometric functions obtained in the case of a second-degree potential.

Carl Gustav Jacobi pointed out, as a surprising property of mathematics, that "one and the same function describes both the true law of pendulum oscillations and the number of representations of an integer as a sum of squares," having in mind precisely this elliptic function.

The elliptic function X is doubly periodic in the real plane of the complex variable t: there exist two complex numbers α and β with a nonreal ratio such that

$$X(t+\alpha) = X(t), \quad X(t+\beta) = X(t)$$

for every $t \in \mathbb{C}$.

Problem. A point mass moves along a line in a field whose potential energy is a polynomial of degree 4 with two local minima.

At which of these potential wells is the period of oscillation (under the same total energy E) greater: at the larger or at the smaller?

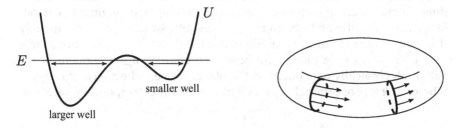

Hint. These periods are the same, because the flows of an incompressible fluid through any two meridians of the torus along which the fluid moves are the same: in unit time, as much fluid enters through one meridian into the region between the two meridians as exits through the other meridian.

The torus to which this needs to be applied is the complex algebraic curve of genus 1 provided by the law of conservation of energy (in the case of a degree-4 potential U, this curve is elliptic).

The general theory of abelian integrals along curves of genus g leads to the following conclusion: if the genus of the algebraic curve M is greater than zero (that is, its Riemann surface is not a sphere), then there exists a rational function R such that the corresponding abelian integral along M cannot be expressed in terms of elementary functions. Indeed, most rational functions have this property, although there are exceptions (for example, $R = 0$).

Probably the first person to consider this topology of abelian integrals with the help of their Riemann surfaces was Newton, who (in connection with Kepler's "law of equal areas" in the theory of planetary motion) considered the following question: *Does there exist a smooth closed curve in the real Euclidean plane such that the area of the segment cut out by a line from the region bounded by the curve is an algebraic function (of the coefficients of the equation of the line)?*

Newton suggested that had such a curve existed, then God would have chosen it for the planetary orbits, rather than the ellipse (so as to make "Kepler's equation"—which determines the position of a planet in its orbit as a function of time—easier to solve for astronomers: for elliptic orbits this equation is not algebraic, but transcendental).

Using the topology of Riemann surfaces, Newton proved that such curves (with algebraic dependence of the area on the secant line) do not exist.

On reading Newton, Leibniz made the notation "ERROR" in the margin of his book, believing that the triangle was a counterexample to Newton's claim.

But Newton's proof was correct: the point is that the triangle is not a smooth curve; its complexification (and Newton used the complex topology of Riemann surfaces in his proof) does not lead to a connected Riemann surface, as in the case considered by Newton, but rather to a "reducible algebraic curve" from three components not forming a single smooth surface.

I further point out that the analogue of Newton's question in three-dimensional space (when one investigates volumes of segments cut out by planes from the body bounded by a smooth surface) has a completely different answer (found by Archimedes): the area cut off from the sphere $x^2 + y^2 + z^2 = 1$ by a plane (and hence the volume of the segment of the ball) is algebraically dependent on the plane. Namely, the area of the region where $z > h$ is proportional to the height $1 - h$ of the corresponding segment.

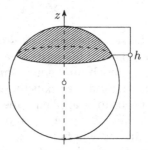

As is clear, ellipsoids (and other second-degree surfaces in Euclidean space $\mathbb{R}^3$) also enjoy this property of algebraicity of volume. But it is not known whether there are other smooth surfaces with this property.

Newton's problem on the volumes of segments in Euclidean spaces $\mathbb{R}^n$ of even dimension n, as in the case of the plane ($n = 2$), leads to the conclusion that there are no cases in which the volume is algebraic (as a consequence of the topology of the corresponding Riemann surfaces). In the case of odd dimension n, as in the three-dimensional case, integrability occurs (for ellipsoids), but it has not been proved whether integrability occurs only for these.[26]

Problem. Investigate the complexification of the two-dimensional sphere.

Here we are dealing with an algebraic surface defined in the affine space $\mathbb{R}^3$ by the equation

$$x^2 + y^2 + z^2 = 1.$$

If one treats x, y, z as complex numbers, then in the space $\mathbb{C}^3$, this equation defines a two-dimensional complex manifold $M^4 \subset \mathbb{C}^3$ (of real dimension 4).

By adding the points at infinity to M^4, we obtain a (smooth) complex algebraic manifold $X^4 \subset \mathbb{C}P^3$.

The problem consists in the (topological) investigation of this compact 4-dimensional manifold.

Answer. $X^4 \approx S^2 \times S^2$ (where the diffeomorphism $\approx$ can even be considered to be holomorphic).

Chapter 6
A Problem for School Pupils

In conclusion, I want to state a topological problem of real projective algebraic geometry that is accessible to preschool children.*

Consider n lines in the real plane $\mathbb{R}^2$. These lines divide the plane into a certain number of regions.

Problem. Into how many regions can n lines divide the plane?

Example. One line divides the plane into two parts. Two parallel lines divide it into three parts, and two nonparallel lines divide it into four parts.

For larger values of n one can easily construct examples of division into M parts, where M has the following values:

n	1	2	3	4	6	7	8
M	2	3	4	5	7	8	9
		4	6	8	12	14	16
			7	9	15	18	22
				10	16	19	?
				11	17	⋮	24
					⋮	29	⋮
					22		37

A few simple examples are as follows:

---

* I point out, by the way, that other problems of real algebraic geometry close to this problem may prove to be algorithmically unsolvable; this relates, for example, to the description of triangulations of tori (and even the 2-dimensional torus) provided by the theory of multidimensional periodic continued fractions. The triangulation ⊠ corresponds to the simplest cubic irrationality, but there is no description of all triangulations corresponding to other cubic irrationalities. The question whether a given triangulation is realizable may prove to be unsolvable.

V.I. Arnold, *Real Algebraic Geometry*, UNITEXT – La Matematica per il 3+2 66,
DOI 10.1007/978-3-642-36243-9_6, © Springer-Verlag Berlin Heidelberg 2013

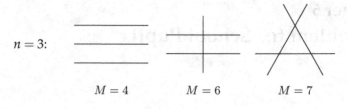

$n = 3$:

$M = 4$ $M = 6$ $M = 7$

$n = 4$:

$M = 5$ $M = 8$ $M = 9$ $M = 10$ $M = 11$

It is perfectly clear that the number M of regions into which n lines divide the plane cannot be less than $n + 1$, that it attains its maximal value $1 + (1 + 2 + \cdots + n) = 1 + n(n+1)/2$, and that it cannot exceed that number.

But are all the intermediate values of M attainable, namely $n + 1 < M < 1 + n(n+1)/2$?

As we shall now prove, not all intermediate values of M are attainable. For example, three lines cannot divide the plane into five parts, and four lines cannot divide the plane into six or seven parts.

Theorem. *The plane cannot be divided into M parts by n (distinct) lines if*

$$n + 1 < M < 2n,$$

nor if

$$2n < M < 3n - 3.$$

For the proof we introduce the following notation. Denote by k the greatest number of parallel lines in the configuration of n lines under consideration. Denote by x the greatest number of lines passing through one point in the configuration of n lines under consideration.

Assertion 1. *If $k = n$, then the number of regions is $M = n + 1$.* (This is obvious; see the first diagrams for $n = 3$ and 4 above.)

Assertion 2. *If $k = n - 1$, then the number of regions is $M = 2n$.*

This too is clear from the above diagrams for $n = 3$ and 4 (the second figure in both cases).

Assertion 3. *If $k = n - 2$, then the number of regions is either $M = 3n - 2$ or $M = 3n - 3$.*

Proof of Assertion 3. A system of $n - 2$ parallel lines and one line not parallel to them divides the plane into $2(n - 1)$ parts (by Assertion 2).

The remaining line intersects our $n - 2$ parallel lines in $n - 2$ different points. If it is parallel to the first line (which is not parallel to the others), then $M = 3(n - 1)$:

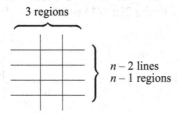

If it is not parallel to the first line and intersects it outside our $n - 2$ parallel lines, then $M = 2(n - 1) + n = 3n - 2$: the addition of this last line produces a further n regions to the $2(n - 1)$ regions that we already had before this extra line was added:

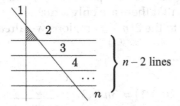

If, on the other hand, the last line intersects the first line in one of its points of intersection with parallel lines, then there is one region fewer; $M = 3n - 3$ (the shaded triangle in the previous diagram disappears):

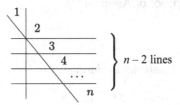

This completes the proof of Assertion 3.

Assertion 4. *If $x = n$, then the number of regions is $M = 2n$ (these n lines divide the plane into $2n$ sectors).*

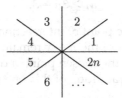

Assertion 5. *If $x = n - 1$, then the number of regions is $M = 3n - 2$ or $M = 3n - 3$.*

Proof of Assertion 5. If the line not passing through the point of intersection of the remaining lines is not parallel to any of them, then it is divided by the $n - 1$ points of its intersection with them into n parts and therefore adds a further n regions to the $2(n - 1)$ regions we already had:

$$M = 2(n - 1) + n = 3n - 2$$

If, on the other hand, the line not passing through the point of intersection of the remaining $n - 1$ lines is parallel to one of them, then the number of its points of intersection with them is equal to $n - 2$, and these points divide the line into $n - 1$ parts, so that there are only a further $n - 1$ regions (instead of n regions) to be added to the $2(n - 1)$ regions we already had:

$$M = 2(n - 1) + (n - 1) = 3n - 3$$

This completes the proof of Assertion 5.

Assertion 6. *If*
$$k < n - 2, \quad 3 < x < n - 1,$$
then the number of regions satisfies the inequality
$$M \geq 4n - 8,$$
while if only
$$k < n - 2, \quad 2 < x < n - 1,$$
then
$$M \geq 3n - 3.$$

Proof of Assertion 6. At the point of intersection of the x lines we consider the $2x$ sectors formed. Each of the $n - x$ remaining lines intersects at least $x - 1$ lines passing through the chosen point and is divided by them into at least x parts. Therefore, the addition of this line gives rise to at least x new regions.

In adding all the $n - x$ remaining lines one by one, we add at least $x(n - x)$ new regions to the $2x$ sectors we already had. Thus the total number of regions will be

$$M \geq 2x + x(n - x) = x(N - x), \quad \text{where } N = n + 2.$$

If, as assumed in Assertion 6, the inequalities $4 \leq x \leq n - 2$ hold, then we also have the inequalities $4 \leq N - x \leq n - 2$, so that the total number M of regions will satisfy the inequality

$$M \geq 4(n - 2) = 4n - 8.$$

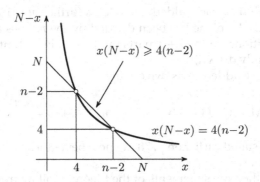

This completes the proof of the first part of Assertion 6.

If the condition $4 \leq x \leq n - 2$ is replaced by $3 \leq x \leq n - 2$, then the pair $(x, N - x)$ can take one further value, $x = 3$, $N - x = n + 2 - 3 = n - 1$, and we obtain (using the same arguments) the following estimate for the number of regions:

$$M \geq 3(n - 1).$$

Assertion 7. *If $x = 2$ (that is, no three lines of the configuration are concurrent) and $2 \leq k \leq n - 3$, then the number of regions satisfies the inequality $M \geq 3(n - 1)$.*

Proof of Assertion 7. Consider the set of k parallel lines of our system of n lines. Each of the $n - k$ remaining lines is divided by these k parallel lines into $k + 1$ parts. Therefore, on drawing the remaining $n - k$ lines, we add to the $k + 1$ parts (into which our k parallel lines divide the plane) at least a further $(n - k)(k + 1)$ parts.

Altogether, we obtain for the total number M of parts the estimate

$$M \geq (k + 1) + (n - k)(k + 1) = (k + 1)(N - (k + 1)),$$

where $N = n + 2$.

Under the conditions of Assertion 7 we have

$$3 \leq k + 1 \leq n - 2, \quad 4 \leq N - (k + 1) \leq N - 3 = n - 1.$$

Hence (as in the proof of Assertion 6) we have the inequality

$$M \geq (k+1)(N - (k+1)) \geq 3(n-1),$$

which completes the proof of Assertion 7.

Assertion 8. *If $x = 2$ and $k = 1$ (so that among our lines there is no pair of parallel lines and no triple of concurrent lines), then for a given n, the maximum possible number M of regions is*

$$M = 1 + \frac{n(n+1)}{2}.$$

Proof of Assertion 8. Each added line adds a further $m + 1$ parts to the regions into which the plane has been divided by the m lines already drawn (since it is partitioned into $m + 1$ parts by its m points of intersection with the m lines already drawn).

Summing these added parts, we obtain

$$M = 1 + (1 + 2 + 3 + \cdots + n) = 1 + \frac{n(n+1)}{2}.$$

The theorem stated earlier on p. 78 can now be deduced from Assertions 1 through 8 as follows.

If $k \geq n - 2$, then the statements of the theorem follow from (the already proved) Assertions 1, 2, and 3.

If $x > n - 2$, then the statements of the theorem follow from Assertions 4 and 5.

If $k \leq n - 3$ with $3 < x < n - 1$, then the first part of Assertion 6 yields the inequality $M \geq 4n - 8$. This number is greater than or equal to $3n - 3$ if $n > 5$, so that the statement of the theorem is true also in this case (for $n \geq 5$).

For $n < 5$ it is also true (and is proved in essence using the above diagrams of systems of three or four lines).

If $k \leq n - 3$ with $x = 3$, then the second part of Assertion 6 yields the statement of the theorem.

If $x = 2$, then for $2 \leq k \leq n - 3$ the statement of the theorem follows from Assertion 7.

In the case $x = 2$, $k = 1$ (which is the only case not yet worked out) Assertion 8 gives the value

$$M = 1 + \frac{n(n+1)}{2}.$$

This value is greater than or equal to $3n - 2$ when $n \geq 3$, so that the inequality of the theorem is proved in this case as well.

Over the intervals

$$n + 1 < M < 2n, \quad 2n < M < 3n - 3,$$

where there are insufficient regions into which n lines divide the plane, there are probably other similar intervals (such as, possibly, $3n - 2 < M < 4n - 8$), but this has not been proved.

I have given an account of this elementary problem expecting the help of school pupils in its solution. I hope that readers will publish this solution.

Recently, A. B. Givental translated into English (or American) Kiselev's remarkable textbook *Geometry*.

In April 2007, in Berkeley, reading the translation of this book, which was well known to me, I discovered that I could not solve one of the problems (although I had solved them all in my youth).

When I compared the two versions, it turned out that this problem was not in Kiselev's original book: it had been added by the translator. The statement of this problem is as follows:

How many lines must one take so as to divide the plane into five parts?

Naturally (since I had arrived in Berkeley from France), I immediately generalized this problem by considering an arbitrary number of lines and parts.

This is how the theorem of this chapter arose, which, as it seems to me, opens up a whole field of activity for school pupils who love mathematics.

I add one further strange circumstance, which was discovered as a result of Givental's translation: he compared Kiselev's textbook *Geometry* with all the others and found in the world literature only one textbook that was as good: this was a book by Hadamard. There arose the conjecture that Kiselev's book was as good as it was thanks to the fact that he had made use of the textbook of that great French mathematician. But a more careful analysis showed that this was not so: Kiselev's book had been published ten years earlier than Hadamard's book. Furthermore, secondary school teachers who used Hadamard's book corrected numerous errors that were initially in it. I hope that my elementary textbook will also be corrected in the same way (and even by school pupils rather than teachers).

Appendix A*
Into How Many Parts Do n Lines Divide the Plane?

Consider n distinct lines in the real projective plane. They divide it into (convex) parts. The question is this: how many parts can be obtained (under all possible arrangements of the lines)?

For small n, the answer is clear; the possible number M of parts is given by the following table:

n	1	2	3	4	5
M	1	2	3	4	5
			4	6	8
				7	9
					10
					11

If we consider one of the lines to be the line at infinity, then we obtain $n-1$ lines in $\mathbb{R}^2$. The values of M in the table are provided by the following configurations of $n-1$ lines:

$n = 2:$ $M = 2$

$n = 3:$ $M = 3$ $M = 4$

$n = 4:$ $M = 4$ $M = 6$ $M = 7$

$n = 5:$ $M = 5$ $M = 8$ $M = 9$ $M = 10$ $M = 11$

---

* This appendix is a translation of the author's article in *Mat. Prosveshchenie*, Ser. 3 12 (2008), 95–104. It was not part of the original Russian edition of this book. It has been appended here at the author's request.

The number of connected components of the complement of the collection of lines in the affine plane $\mathbb{R}^2$ is studied in similar fashion. This is the same problem, since one can declare one of the n lines to be the line at infinity and study the complement of the $n-1$ lines in the affine plane (which coincides with the complement of the n lines in the projective plane).

Looking at the previous examples, we observe that the smallest number of parts of the complement of n lines in $\mathbb{R}P^2$ is $M = n$, while the largest is

$$M = 1 + (1 + 2 + 3 + \cdots + (n-1)) = 1 + \frac{n(n-1)}{2}.$$

However, only some (and not all) of the intermediate numbers between these limits can be attained. Starting from some point, all sufficiently large values of M are attained, but the beginning of the list of attainable values contains gaps (or holes).

The aim of this paper is to describe these holes. Here is the first hole.

Theorem 1. *The value $M = 2(n-1)$ is attainable, but no other value of M in the interval*

$$n < M < 2(n-1)$$

is attainable.

For any choice of $n > 2$ lines in $\mathbb{R}P^2$, their complement cannot consist of such a number M of connected components.

Proof. Denote by k the greatest number of lines (among our n lines) passing through one point.

Lemma 1. *If $k = n$, then $M = n$.*

Proof. We assume that one of these n lines is the line at infinity. Then the remaining lines are parallel. They divide the affine plane $\mathbb{R}^2$ (complementary to the first line) into n parts, since there are $n-1$ such parallel lines.

Lemma 2. *If $k = n - 1$, then $M = 2(n-1)$.*

Proof. We assume that the remaining nth line is the line at infinity. Its complement is $\mathbb{R}^2$. The family of $n-1$ lines passing through one point divides the plane $\mathbb{R}^2$ into $2(n-1)$ parts, as required.

Lemma 3. *If $k \leq n - 1$, then $M \geq 2(n-1)$.*

Proof. Note that for $n > 1$ we have $k \geq 2$ (since any two lines in $\mathbb{R}P^2$ intersect). We choose one of the k lines passing through one point as the line at infinity. Its complement is the affine plane $\mathbb{R}^2$, which contains $k-1$ parallel lines of the chosen pencil together with a further $n - k \geq 1$ remaining lines.

These parallel lines divide the plane $\mathbb{R}P^2$ into k parts. By adding the remaining lines one by one, we shall gradually increase the number of parts. Here, if the sth line to be added is intersected by the lines already existing in

x_s points, then it is divided by them into x_s segments, each of which divides one of the parts that existed before the sth line was added in two. Therefore, when the sth line is added, the number of parts of the complement is increased by exactly x_s.

We now observe that $x_s \geq k$ (since the line being added intersects all k parallel lines of the original pencil in k distinct points). Therefore the total number of added parts is

$$x_1 + \cdots + x_{n-k} \geq k(n-k).$$

By adding these parts to the k parts that we already had before dealing with the $n - k$ "additional" lines, we conclude that

$$M \geq k + k(n-k) = k(n-k+1). \tag{1}$$

The sum of both factors on the right-hand side is equal to $n + 1$. When two positive numbers whose sum is equal to $n + 1$ are multiplied together, the greater the smaller factor is, the greater their product:

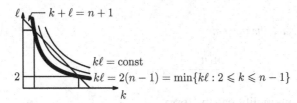

If $2 \leq k \leq n - 1$, then

$$\min_{k+l=n+1} (k,l) \geq 2,$$

so that

$$M \geq 2(n-2+1) = 2(n-1).$$

This completes the proof of Lemma 3

Theorem 1 now follows from Lemmas 1, 2, and 3 (since any number $k \leq n$ is either equal to n or equal to $n - 1$ or less than $n - 1$).

The first hole has now been described. Next we describe the second hole. Let $n \geq 3$.

Theorem 2. *The value $M = 3n - 6$ is attainable, but no other value of M in the interval*

$$2(n-1) < M < 3(n-2)$$

is attainable.

For any choice of $n > 2$ lines in $\mathbb{RP}^2$, their complement cannot consist of such a number M of connected components.

Proof. As before, we denote by k the maximum number of lines (among the given n lines) passing through one point, and we call one of them the line at infinity. We regard the remaining $n-1$ lines in the affine plane $\mathbb{R}^2$, which is the complement of the chosen line, as the pencil of $k-1$ parallel lines supplemented by the $n-k$ remaining lines not parallel to these $k-1$ lines of the pencil.

If $k = n-1$, then $M = 2(n-1)$ by Lemma 2. If, on the other hand, $k \leq n-2$, then from formula (1) in the proof of Lemma 3 we obtain

$$M \geq k(n-k+1) \geq (n-2)((n+1)-(n-2)) = 3(n-2),$$

provided that $k \geq 3$ (for which $\min_{k+l=n+1}(k,l) \geq 3$).

Thus, the theorem is proved for all arrangements of n lines such that $k > 2$. We now prove that $M \geq 3(n-2)$ in the remaining case, in which $k = 2$.

If $k = 2$, that is, no three lines are concurrent, then all our n lines divide the plane into the maximum possible number of parts for n lines (and attainable for n lines in general position), namely

$$M = 1 + \frac{n(n-1)}{2}.$$

Lemma. *We have the inequality*

$$\frac{n(n-1)}{2} + 1 \geq 3(n-2).$$

Proof. This inequality has the form

$$n^2 - n - 6(n-2) + 2 \geq 0,$$

that is,

$$n^2 - 7n + 14 \geq 0,$$

which is true because the discriminant of the quadratic trinomial on the left-hand side,

$$49 - 4 \cdot 14,$$

is negative.

Thus the lemma is proved, so that the inequality $M \geq 3(n-2)$ holds in the case $k = 2$ as well.

Having proved Theorem 2, we have described the second hole. It first appears when $n = 6$ (when in the interval defining the hole there are integer points: $3(n-2) - 2(n-1) > 1$ for $n > 5$).

The subsequent holes that we now study are under the assumption that the number n of lines is sufficiently large (in comparison with the number of the hole).

Theorem 3. *Suppose that the greatest number of lines (among our n lines) passing through one point is k. Then these n lines divide the plane $\mathbb{R}P^2$ into M parts, where M lies in the interval*

$$k(n+1-k) \leq M \leq k(n+1-k) + \frac{r(r-1)}{2}, \quad \text{where } r = n - k.$$

(Here all the numbers M in this interval are attainable for a suitable choice of n lines, provided that n is sufficiently large.)

Proof. Choose a pencil of k lines. The lines of this pencil divide the plane $\mathbb{R}P^2$ into k parts. The remaining $n - k = r$ lines add to k the following number of parts:

$$M' = x_1 + x_2 + \cdots + x_{n-k},$$

where x_s is the number of points of the sth added line intersected by the previous lines.

Among these previous lines there are k lines of the chosen pencil (and $s - 1$ added lines). The points of intersection with the lines of the pencil are all distinct (since the only common point of two lines of the pencil is the point of intersection of the k lines of the pencil originally chosen, so that no other of our n lines can pass through this point).

Hence $x_s \geq k$, $M' \geq k(n - k)$, $M \geq k(n - k + 1)$, which proves the first inequality in Theorem 3.

On the other hand, $x_s \leq k + (s - 1)$. Consequently,

$$M' \leq k(n-k) + (0+1+\cdots+n-k-1) = k(n-k) + \frac{r(r-1)}{2},$$

$$M \leq k(n+1-k) + \frac{r(r-1)}{2}.$$

This proves the second inequality in Theorem 3.

All the values of M in the interval described by both inequalities in the above theorem are attained (for sufficiently large n) for the following reason.

The greatest number of parts is provided by choosing the r additional lines in general position. For them all the points of intersection (with the lines of the pencil and with each other) are distinct; as a result, this gives us $k(n+1-k) + r(r-1)/2$ parts.

If n is sufficiently large in comparison with r (for example,[27] if $n \geq r$ $(r+1)/2$), then one can choose the additional lines so that any chosen points in the pairwise intersection of the lines lie on the lines of the pencil (so that $x_s = k$ for the corresponding s).

In fact, we can, for example, start with r lines in general position in the affine plane $\mathbb{R}^2$ and draw, through any collection of $r(r-1)/2 - S$ points of their pairwise intersection, lines that are parallel to each other but not parallel to these r lines. By including these parallel lines along with the line at infinity in the pencil of $k = n - r$ parallel lines, we obtain a collection of lines for which M', the sum of x_s over all $s = 1,\ldots,r$, exceeds kr precisely by S. In this case,[28] $M' = kr + S$, $M = k(n - k + 1) + S$, and Theorem 3 is proved.

Theorem 4. *Suppose that the greatest number of lines (among our n lines) passing through one point is $k > 2$. Then these n lines divide the plane $\mathbb{R}P^2$ into M parts, where*

$$M \geq \frac{n(n-1)}{2(k-1)}.$$

Here it is important that the numerator increases with the number n of lines at the same rate as n^2, while the denominator is independent of the number n. As a result, the right-hand side becomes larger than any linear function of n for sufficiently large n (when k is fixed).

To prove Theorem 4 we arrange the given n lines in some order. By an "event" we mean the intersection of some line with the lowest-numbered line. Thus, the number of events is $0 + 1 + 2 + \cdots + (n - 1) = n(n-1)/2$ (independently of how many distinct points of intersection there are).

By a "partition" we mean a division into parts of some line (say the sth) by lines with lower numberings. We denote by x_s the number of points in the partition of the sth line. These x_s points divide that projective line into x_s parts.

We add the lines one by one, and at each stage, we increase the number of components of the complement of the lines by the number x_s of parts added by the sth line (which divides each of the already existing x_s components intersected by it into two parts).

Therefore the total number of components of the complement of the union of the n lines in the projective plane $\mathbb{R}P^2$ is given by

$$M = \sum_{s=1}^{n} x_s,$$

supposing formally that $x_1 = 1$: although the first line does not divide any "preceding" lines, one needs to take into account the (single) component of the complement of one line in the *projective* plane.

At each point of the partition, the maximum number of occurring events is $k - 1$ (the intersection of the sth line with the preceding ones), since there cannot be more than k lines of our collection passing through one point. Therefore, *the number of all events does not exceed $M(k - 1)$*. And since the latter is equal to $n(n - 1)/2$, we conclude that

$$\frac{n(n-1)}{2} \leq M(k-1),$$

that is,

$$M \geq \frac{n(n-1)}{2(k-1)},$$

which completes the proof of Theorem 4.

For the investigation of "stable" holes (the jth stable hole D_j will be investigated under the assumption that the number n exceeds some constant depending on j), we introduce the following notation:

$$\alpha_j = (n-j)(j+1), \quad \beta_j = \frac{(n-j)(j+1)+j(j-1)}{2}.$$

For sufficiently large n, the first terms of these two sequences are arranged in the following order:

$$\alpha_0 = \beta_0 < \alpha_1 = \beta_1 < \alpha_2 < \beta_2 < \alpha_3 < \beta_3 < \cdots < \alpha_{j-1} < \beta_{j-1} < \alpha_j.$$

We denote by $P_0, P_1, \ldots$ the closed intervals

$$P_0 = [\alpha_0 \le M \le \beta_0], \quad P_j = [\alpha_1 \le M \le \beta_1], \quad \ldots, \quad P_1 = [\alpha_j \le M \le \beta_j],$$

and by $D_1, D_2, \ldots, D_j$ the complementary open intervals

$$D_1 =]\beta_0 < M < \alpha_1[, \quad D_2 =]\beta_1 < M < \alpha_2[, \quad \ldots, \quad D_j =]\beta_{j-1} < M < \alpha_j[.$$

The stable hole D_j is described as follows.

Theorem 5. *If the number of lines n is sufficiently large, then the number M of components of their complement in the projective plane $\mathbb{RP}^2$ cannot take values in the intervals D_j. In other words, no value of M for which*

$$\beta_{j-1} = j(n+1-j) + \frac{(j-1)(j-2)}{2} < M < (j+1)(n-j) = \alpha_j$$

is possible.

Proof. Denote by k the greatest number of lines (among our n lines) passing through one point. We shall prove that M cannot lie in an interval D_j for any k, but this proof will be based on different considerations in the following three cases:

I. $k > n - j$;

II. II $j + 1 \le k \le n - j$;

III. $k \le j$.

Here we suppose that $n - j \ge j + 1$ (which holds for n sufficiently large).

Case I. Suppose that k takes one of the values $\{n, n-1, \ldots, n-j+1\}$, and put $r = n - k$.[29]

According to Theorem 3, the number M lies in the interval

$$\alpha_r = (n-r)(r+1) \le M \le (n-r)(r+1) + r(r-1)/2 = \beta_r,$$

that is, $M \in P_r, 0 \le r \le j-1$.

None of these r segments intersects the interval D_j, so Theorem 5 is proved for case I.

Case II. Suppose that $j+1 \le k \le n-j$. Then $\ell = n+1-k$ also satisfies the inequalities $j+1 \le \ell \le n-j$. In this case, $\min_{k+\ell=n+1}(k,\ell) \ge j+1$:

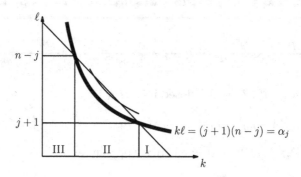

Therefore, again by Theorem 3,

$$M \ge (n-j)(j+1) = \alpha_j.$$

However, $M < \alpha_j$ in the interval D_j. Thus Theorem 5 is proved for Case II as well.

Case III. Suppose that $2 < k \le j$. According to Theorem 4,

$$M \ge \frac{n(n-1)}{2(k-1)} \ge \frac{n(n-1)}{2(j-1)}.$$

The right-hand side of this inequality is greater than α_j if n is sufficiently large. Indeed,

$$\frac{n(n-1)}{2(j-1)} \ge (n-j)(j+1)$$

for sufficiently large n, because then

$$\frac{n(n-1)}{n-j} > 2(j^2 - 1).$$

For example,

$$\frac{n(n-1)}{n-j} \ge n,$$

so that the condition $n > 2(j^2 - 1)$ suffices for the inequailty $M > \alpha_j$, which does not allow M to lie in the interval D_j (between β_{j-1} and α_j).

In the only case not yet disposed of, $k = 2$, our n lines in general position divide the projective plane $\mathbb{R}P^2$ into $M = 1 + n(n-1)/2$ parts.

This number M is greater than the limit $\alpha_j = (n-j)(j+1)$, since for $j + 1 < n/2$ (which we have supposed), the inequalities

$$\alpha_j < \frac{n(n-1)}{2} < M$$

hold for $j > 0$.

Thus Theorem 5 is proved for $k = 2$ as well, and it is therefore proved for all k (the case $k = 1$ for $n > 1$ is not realized, since every pair of lines in the projective plane intersect).

Thus Theorem 5 is completely proved, so that (for a sufficiently large number of lines n) all the stable holes

$$D_1, D_2, \ldots, D_j, \ldots$$

exist in the sequence of numbers M of components into which n lines divide the real projective plane.

Remark 1. I do not know whether the unstable holes (for values of n smaller than those indicated above) are given by the same formulas ($\beta_{j-1} < M < \alpha_j$) as the stable holes. To begin with, instability does not manifest itself for small j.

The first unclear case is the third hole for $n = 9$. In this case, the formulas give $\alpha_3 = 24$, $\beta_2 = 22$.

Nine lines can divide the projective plane into 22 regions and into 24 regions. Whether they can divide it into 23 regions or, on the contrary, whether $M = 23$ for $n = 9$ is the third hole, is unknown. Such an arrangement of lines (if it is possible) would be possible only in the case that no four of these nine lines pass though a single point (case $k = 3$ in the proof of Theorem 5).

Remark 2. The motivation for this article was the publication in Berkeley of A.B. Givental's translation into English of A.P. Kiselev's book *Geometry*. When in April 2007 in California I looked at this translation, I was unable to solve one of the problems in this book (I had resolved them all in my youth).

This problem was as follows: how many lines divide the plane $\mathbb{R}^2$ into five convex parts?

I asked Givental how this question was formulated in Kiselev's original book; he confessed that this was by no means the case: the problem was added by the translator (who improved Kiselev in other places as well).

Every mathematical problem admits two versions: the Russian version, which nobody can simplify (without losing the essence of the problem), and the French version, which nobody can generalize any further (since it has already been formulated in such a general form that it contains all possible generalizations).

On arriving from Paris to Berkeley I decided to formulate the French version of Givental's problem. With this in mind, I changed five regions to any number M of regions. This is how the present article came about.

I have been unable to solve this general problem. One should be able to describe all holes for all values of n; but even the third hole was calculated by me explicitly only for $n \geq 14$ (when it becomes stable). By lecturing local school pupils in Berkeley, Stanford, San José, and Santa Clara (where heroic leaders of the Moscow-style mathematical circles have taught the pupils how to solve difficult problems better than I), I hoped that they would come up with a description of the unstable holes, which is still awaited.

It seems that the question whether 9 lines can divide the projective plane into 23 regions remains unsolved.[30]

Editors' Comments on Gudkov's Conjecture

It is difficult to overestimate Arnold's role in the revolutionary explosive developments in the study of topological properties of real algebraic varieties (and in particular, in solving Hilbert's sixteenth problem). A whole series of works was inspired by Arnold's foundational paper "On the Arrangements of Ovals of Real Plane Algebraic Curves, Involutions of Four-Dimensional Smooth Manifolds, and the Arithmetic of Integer Quadratic Forms" (*Funk. Anal. i Prilozh.* 5 (1971), no. 3, 1–9). The very title of this important paper indicated a new direction of investigation in this field. A number of new results in this paper were immediately named after Arnold, such as Arnold's congruence, Arnold's inequalities, and so on.

Arnold's paper was stimulated by (private and public) communications with D.A. Gudkov and V.A. Rokhlin. Gudkov (in his 1971 note in *Doklady AN SSSR* 200 (1971), no. 6, 1269–1272, and a 1970 talk in Moscow) had formulated as a conjecture the congruence modulo 8 for maximal curves* of even degree. Rokhlin (in his 1971 paper *Funk. Anal. i Prilozh.* 5 (1971), no. 1, 48–60) had obtained his genus bounds in 4-dimensional topology by means of the Atiyah–Singer–Hirzebruch formula for signatures of ramified coverings (he presented these results at a 1970 talk in Moscow; it may also be worth mentioning a slightly earlier talk given by Rokhlin in Moscow about his attempts to prove the Whitney conjecture using congruences modulo 16). Arnold came to his foundational discoveries through his detailed knowledge of Gudkov's habilitation thesis, in which Gudkov had completed the classification of real nonsingular plane projective curves of degree 6, and his awareness of both Gudkov's conjecture and Rokhlin's results. (Arnold himself, in the abstract of his talk given at the Moscow Mathematical Society meeting of April 6, 1971, mentioned both Gudkov's authorship in stating the conjecture and the role of Rokhlin's results on signatures of ramified coverings.)

---

* A *maximal curve* (also called an *M-curve*) is a smooth real algebraic curve whose number of real components is equal to $g + 1$, where g is the genus of the curve.

V.I. Arnold, *Real Algebraic Geometry*, UNITEXT – La Matematica per il 3+2 66,
DOI 10.1007/978-3-642-36243-9, © Springer-Verlag Berlin Heidelberg 2013

As to the history of Gudkov's conjecture, Rokhlin told us, and Gudkov confirmed, the following story (letters corroborating this story were found by G. M. Polotovsky in Gudkov's personal archive). During the preparation of his habilitation thesis, Gudkov showed a preliminary version of part of it to V. V. Morozov, a professor at Kazan University. This version contained detailed proofs of a complete classification of smooth degree-6 curves, which, as was discovered later, contained a few errors at this stage. Morosov (who later was one of the official referees of Gudkov's habilitation) pointed out to Gudkov a strange—and doubtful in Morosov's opinion—asymmetry in a diagram expressing the classification obtained. It is in rectifying this asymmetry that Gudkov came to the correct classification and thence to his conjecture.

The first traces of the conjecture (in its complete form) are found in a letter of Gudkov to Morosov (winter 1969–1970, Gudkov's archive, communicated by Polotovsky), in the verbatim record of the habilitation defense (1970, Gudkov's archive, communicated by Polotovsky), and in his paper "Construction of New Series of M-Curves," *Doklady AN SSSR 200* (1971), no. 6, 1269–1272.

Additional documentary confirmation that Gudkov was the author of the conjecture for M-curves is a letter from Gudkov to Arnold dated October 15, 1972 (Gudkov's personal archive, communicated by Polotovsky), in which Gudkov proposes an additional conjecture and writes to Arnold that "*I already had this in mind when I formulated the conjecture for M-curves, but perfidiously concealed it from you and Rokhlin.*"

Let us conclude by observing that as Arnold wrote in "On the Arrangements of Ovals of Real Plane Algebraic Curves, Involutions of Four-Dimensional Smooth Manifolds, and the Arithmetic of Integer Quadratic Forms," that paper would not exist if Gudkov had not communicated his conjecture to Arnold.

Notes

1. Questions concerning the topology of real algebraic varieties form the first part of Hilbert's sixteenth problem. Below we present an English translation of the text of this part. (Hilbert's lecture was translated by Mary Winston Newson for the *Bulletin of the American Mathematical Society* 8 (1902), 437–479.) It is exactly this part, under the title "Problem der Topologie algebraischer Curven und Flächen," that was presented by Hilbert in 1900 as the sixteenth problem in his renowned lecture at the second International Congress of Mathematicians. For the lecture he selected 10 problems. The famous list of 23 problems appeared in its final form only later and was published in 1901 in *Archiv der Mathematik und Physik*:

> The maximum number of closed and separate branches which a plane algebraic curve of the *n*th order can have has been determined by Harnack (*Mathematische Annalen* 10). There arises the further question as to the relative position of the branches in the plane.
>
> As to curves of the 6th order, I have satisfied myself—by a complicated process, it is true—that of the eleven branches which they can have according to Harnack, by no means all can lie external to one another, but that one branch must exist in whose interior one branch and in whose exterior nine branches lie, or inversely. A thorough investigation of the relative position of the separate branches when their number is the maximum seems to me to be of very great interest, and not less so the corresponding investigation as to the number, form, and position of the sheets of an algebraic surface in space. Till now, indeed, it is not even known what is the maximum number of sheets which a surface of the 4th order in three-dimensional space can really have. (cf. Rohn, "Flächen vierter Ordnung," *Preisschriften der Fürstlich Jablonowskischen Gesellschaft*, Leipzig 1886).

Hilbert's questions on curves of degree 6 in the real projective plane and surfaces of degree 4 in three-dimensional real projective space have found their complete answers through the work of D. A. Gudkov, V. I. Arnold, V. A. Rokhlin, and V. M. Kharlamov in the breakthrough of 1969–1976; see http://www.pdmi.ras.ru/~olegviro/H16-e.pdf for a detailed discussion.

2. In principle, one could add to this list some other real forms of curves described by a second-degree equation, namely pairs of intersecting complex conjugate imaginary lines (such as the pair given by the equation $x^2 + y^2 = 0$), pairs of parallel complex conjugate imaginary lines (such as the pair given by the equation $x^2 + 1 = 0$), and empty "ellipses" (such as the one given by the equation $x^2 + y^2 + 1 = 0$). A possible reason not to include them in the list is the fact that they do not represent a curve in the usual purely real sense, but either a point or the empty set.

3. Arnold is not very precise in this comment. Perhaps he is intentionally forcing his readers to think further and recall that every point that can be constructed with straightedge and compass can be also constructed with compass alone.

4. Presumably, Arnold has here in mind the following beautiful lines from Goethe's poem "Gott, Gemüt und Welt" (God, Soul, and World): "Willst du ins Unendliche schreiten; Geh nur im Endlichen nach allen Seiten." This can be translated literally as "If you would step into the infinite, you have only to walk in the finite in all directions." Or more poetically, "If to the Infinite you want to stride, Just walk in the Finite to every side."

5. A *semicubic cusp* (or an *ordinary cusp*) is an isolated singularity that in appropriate local coordinates is given by the equation $y^2 = x^3$.

6. Even if the precise meaning of such a philosophical statement is unclear, it may push us to think about a kind of asymptotic real algebraic geometry. Such a field does not exist yet, but see V. M. Kharlamov, S. Yu. Orevkov, "The Number of Trees Half of Whose Vertices Are Leaves and Asymptotic Enumeration of Plane Real Algebraic Curves," *J.*

Combin. Theory Ser. A 1051 (2004), no. 1, 127–142, where it is shown that the only known asymptotically significant results on the topology of real plane algebraic curves are some consequences of Bézout's theorem and certain Arnold inequalities (and not the Gudkov congruence, contrary to the impression given by Arnold's remarks on p. 45).

7. The discovery of the Möbius strip is usually cited as originating with a paper by Möbius on another subject: "On the Determination of the Volume of a Polyhedron."

Drawings representing such a shape date back to ancient times; it is for, example, evident in the famous Ouroboros drawing from the early alchemical text the *Chrysopoeia of Cleopatra* from the second century.

8. If by Möbius's assertion one understands the statement that for a smooth projective curve obtained from a projective line by a deformation in the class of embedded circles, the number of inflection points is odd and greater than or equal to 3, then such a statement is proved, for example, by S. Sasaki in *Tôhoku Math. J.* 2 (1957), no. 9, 113–117 (another proof and a bit more geometric information can be found in R. Pignoni, *Manuscripta Math* 72 (1991) 223–249). Indeed, even in a stronger form, such an inflection property holds for every embedded smooth curve that is not contractible to a point. In all these statements, it is more convenient to define an inflection point as a point where locally the curve goes from one side of the tangent line to the other.

On the other hand, there is another question, which was raised by Arnold himself and which is still open. Namely, Arnold conjectured a statement that reinforces Möbius's theorem. It is based on the notion of *dangerous self-tangencies* of an immersed circle. Following Arnold, a self-tangency is called dangerous if an orientation of the circle induces, by means of the tangent branches, the same direction on the tangent at the tangency point. The conjecture states that every projective curve obtained from a projective line by deformation in the class of immersed circles without dangerous self-tangencies has at least three inflection points. Some partial results in this direction can be found in D. Panov, *Funct. Analysis and Its Appl.* 32 (1998), no. 1, 23–31.

9. In fact, in the same volume of *Mathematische Annalen* that contains the paper by A. Harnack, there appeared a paper by F. Klein (at that time editor of *Mathematische Annalen*). Harnack's paper was submitted in January, and Klein's in April. Formally, Klein's paper concerns a related but different topic. Klein, recognizing that the topological technique from his paper allowed him to give another proof of Harnack's bound for the number of connected components of a real algebraic curve by $g + 1$ (this gives half of Harnack's theorem; the other half is the realizability of this bound by plane curves in any degree), inserted such a proof as an additional section.

Let us observe also that Harnack's proof of the inequality is different. He never explicitly manipulates the Riemann surface of the curve, using instead Bézout's theorem. As to the genus, it appears in Harnack's proof as the standard expression $(d-1)(d-2)/2$ diminished by the number of nodes and cusps (Harnack does not consider more complicated singularities).

10. This discussion is in fact postponed to Chapter 5.

11. Neither Newton nor Descartes carried out or even intended a "full" investigation of real plane curves of degree 4. Such an investigation was launched by A. Cayley and continued, in what concerns at least the nonsingular quartics, by F. Klein and H.-G. Zeuthen. Stratifications (of the same type as Newton's stratifications in the case of degree 3) of the space of real plane curves of degree 4 were produced in a series of papers by D. A. Gudkov and his collaborators; see references and further information in A. B. Korchagin and D. A. Weinberg, *Rocky Mountain J. Math.* 32 (2002), no. 1, 255–347.

12. This is probably a misprint, and Arnold meant the opposite statement. The topology of nonsingular real curves of degree 5 and 7 is somehow less complicated than that of curves of degree 6. It seems to be a general rule: the real curves of degree $2k + 1$ are, in a sense, not more complicated than the real curves of degree $2k$.

13. This sentence is a bit confusing. The situation was in fact as follows. It is in his PhD thesis that Gudkov "proved" Hilbert's statement on the arrangements of 11 ovals of real

sextics. A little while later, he found a mistake in the proof, and after correcting the mistake, completed the classification of all possible arrangements of nonsingular sextics (even with any number of ovals). This correct classification (which refutes Hilbert's theorem) became the core of Gudkov's habilitation thesis.

14. Gudkov's result on M-sextics excludes only a small part of these billion possibilities. A majority is in fact excluded by Bézout's-theorem arguments. For curves of higher degrees as well, Bézout's-theorem arguments forbid many more oval arrangements than Gudkov's congruence (cf., endnote 6).

15. Arnold's discussion of Gudkov's conjecture may give the impression that it was Arnold and not Gudkov who conjectured the congruence. We refer the reader to the comments on Gudkov's conjecture at the end of this book, where we present a short review that refutes such an attribution of the conjecture.

16. Rokhlin's proof is based on different tools, but it uses the same 4-manifold as in Arnold's proof.

17. At present, 83 configurations have been realized, and 6 remain uncertain; see S. Yu. Orevkov, *Funct. Analysis and Appl.* 36 (2002), 247–249, and *Geometric and Functional Analysis* 12 (2002), no. 4, 723–755. The questionable arrangements are shown in the following figures:

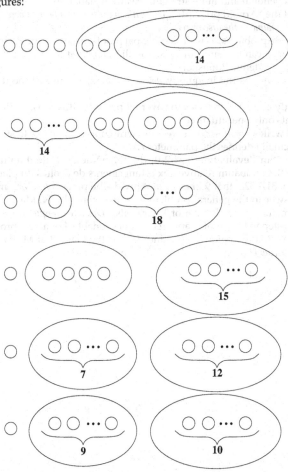

18. The expression "the numbers of points of intersection with lines" means the restrictions that arise from intersecting the curves with lines (as explained a few pages above in the case of quartics), as well as with conics.

19. In what follows, this definition is understood in such a way that it is necessary to assume, in addition, that $g_0 = h_0 = $ id.

20. In an equivalent form, that is, regarding $K(n)$ as the number of up–down sequences, this theorem is an old result of D. André ("Développement de $\sec x$ et de $\tan x$," *C. R. Acad. Sci. Paris* 88 (1879), 965–967).

21. Additional information and references can be found in J. Millar, N. J. A. Sloane, and N. E. Young, *J. Combinatorial Theory, Series A* 76 (1996), 44–54.

22. The lower bound given by Ortiz-Rodriguez was improved by 2 in E. Brugallé and B. Bertrand, *C. R. Math. Acad. Sci. Paris* 348 (2010), no. 5–6, 287–289; compare the next note.

23. This lower bound, similar to that on p. 52, was also improved by 2 in E. Brugallé and B. Bertrand, *C. R. Math. Acad. Sci. Paris* 348 (2010), no. 5–6, 287–289.

24. In fact, it has already been established above that "complex projective circles" are homeomorphic (and even diffeomorphic) to a 2-dimensional sphere, which gives the same result for complex projective lines, since they are diffeomorphic to complex circles. Arnold gives below an independent and more straightforward explanation.

25. This is an illustration of the "Arnold principle": *If a mathematical notion bears a personal name, then that name is not the name of the discoverer.*

26. More information on these problems, as well as principal proofs (based on the Picard–Lefschetz monodromy theory), can be found in V. A. Vassiliev, *Ramified Integrals, Singularities, and Lacunas,* Kluwer Academic Publishers, 1995.

27. In the original, Arnold requires the inequality $n \geq r(r-1)/2$, but in fact, he uses $n \geq r(r+1)/2$.

28. Here the proof is slightly corrected in order to cover all possible situations, while in the original, Arnold presents only one situation.

29. In the original, Arnold writes $k = n - r$, but we have written it as $r = n - k$ to stress the fact that this formula actually defines the parameter r.

30. R. Cordovil showed in "Sur l'evaluation $t(M;2,0)$ du polynôme de Tutte d'un matroïde et une conjecture de B. Grünbaum relative aux arrangements de droites du plan," *European J. Combin.* 1 (1980), 317–322, that 9 lines cannot divide the projective plane into 23 regions. A complete answer to the general problem was discovered by N. Martinov: "Classification of Arrangements by the Number of Their Cells," *Discrete Comput. Geom.* 9 (1993), 39–46. An induction step in his proof is apparently questionable. For a new proof, see I. Shnurnikov, "Into How Many Regions Do n Lines Divide the Plane If at Most k of Them Are Concurrent?" *Moscow Univ. Math. Bull.* 65 (2010), no. 5, 208–212.

Collana Unitext – La Matematica per il 3+2

Series Editors:
A. Quarteroni (Editor-in-Chief)
L. Ambrosio
P. Biscari
C. Ciliberto
G. van der Geer
G. Rinaldi
W.J. Runggaldier

Editor at Springer:
F. Bonadei
francesca.bonadei@springer.com

As of 2004, the books published in the series have been given a volume number. Titles in grey indicate editions out of print.
As of 2011, the series also publishes books in English.

S. Margarita, E. Salinelli
MultiMath – Matematica Multimediale per l'Università
2004, XX+270 pp, ISBN 88-470-0228-1

A. Quarteroni, R. Sacco, F.Saleri
Matematica numerica (2a Ed.)
2000, XIV+448 pp, ISBN 88-470-0077-7
2002, 2004 ristampa riveduta e corretta
(1a edizione 1998, ISBN 88-470-0010-6)

13. A. Quarteroni, F. Saleri
Introduzione al Calcolo Scientifico (2a Ed.)
2004, X+262 pp, ISBN 88-470-0256-7
(1a edizione 2002, ISBN 88-470-0149-8)

14. S. Salsa
Equazioni a derivate parziali - Metodi, modelli e applicazioni
2004, XII+426 pp, ISBN 88-470-0259-1

15. G. Riccardi
Calcolo differenziale ed integrale
2004, XII+314 pp, ISBN 88-470-0285-0

16. M. Impedovo
Matematica generale con il calcolatore
2005, X+526 pp, ISBN 88-470-0258-3

17. L. Formaggia, F. Saleri, A. Veneziani
Applicazioni ed esercizi di modellistica numerica
per problemi differenziali
2005, VIII+396 pp, ISBN 88-470-0257-5

18. S. Salsa, G. Verzini
Equazioni a derivate parziali – Complementi ed esercizi
2005, VIII+406 pp, ISBN 88-470-0260-5
2007, ristampa con modifiche

19. C. Canuto, A. Tabacco
Analisi Matematica I (2a Ed.)
2005, XII+448 pp, ISBN 88-470-0337-7
(1a edizione, 2003, XII+376 pp, ISBN 88-470-0220-6)

20. F. Biagini, M. Campanino
    Elementi di Probabilità e Statistica
    2006, XII+236 pp, ISBN 88-470-0330-X

21. S. Leonesi, C. Toffalori
    Numeri e Crittografia
    2006, VIII+178 pp, ISBN 88-470-0331-8

22. A. Quarteroni, F. Saleri
    Introduzione al Calcolo Scientifico (3a Ed.)
    2006, X+306 pp, ISBN 88-470-0480-2

23. S. Leonesi, C. Toffalori
    Un invito all'Algebra
    2006, XVII+432 pp, ISBN 88-470-0313-X

24. W.M. Baldoni, C. Ciliberto, G.M. Piacentini Cattaneo
    Aritmetica, Crittografia e Codici
    2006, XVI+518 pp, ISBN 88-470-0455-1

25. A. Quarteroni
    Modellistica numerica per problemi differenziali (3a Ed.)
    2006, XIV+452 pp, ISBN 88-470-0493-4
    (1a edizione 2000, ISBN 88-470-0108-0)
    (2a edizione 2003, ISBN 88-470-0203-6)

26. M. Abate, F. Tovena
    Curve e superfici
    2006, XIV+394 pp, ISBN 88-470-0535-3

27. L. Giuzzi
    Codici correttori
    2006, XVI+402 pp, ISBN 88-470-0539-6

28. L. Robbiano
    Algebra lineare
    2007, XVI+210 pp, ISBN 88-470-0446-2

29. E. Rosazza Gianin, C. Sgarra
    Esercizi di finanza matematica
    2007, X+184 pp, ISBN 978-88-470-0610-2

30. A. Machì
Gruppi – Una introduzione a idee e metodi della Teoria dei Gruppi
2007, XII+350 pp, ISBN 978-88-470-0622-5
2010, ristampa con modifiche

31 Y. Biollay, A. Chaabouni, J. Stubbe
Matematica si parte!
A cura di A. Quarteroni
2007, XII+196 pp, ISBN 978-88-470-0675-1

32. M. Manetti
Topologia
2008, XII+298 pp, ISBN 978-88-470-0756-7

33. A. Pascucci
Calcolo stocastico per la finanza
2008, XVI+518 pp, ISBN 978-88-470-0600-3

34. A. Quarteroni, R. Sacco, F. Saleri
Matematica numerica (3a Ed.)
2008, XVI+510 pp, ISBN 978-88-470-0782-6

35. P. Cannarsa, T. D'Aprile
Introduzione alla teoria della misura e all'analisi funzionale
2008, XII+268 pp, ISBN 978-88-470-0701-7

36. A. Quarteroni, F. Saleri
Calcolo scientifico (4a Ed.)
2008, XIV+358 pp, ISBN 978-88-470-0837-3

37. C. Canuto, A. Tabacco
Analisi Matematica I (3a Ed.)
2008, XIV+452 pp, ISBN 978-88-470-0871-3

38. S. Gabelli
Teoria delle Equazioni e Teoria di Galois
2008, XVI+410 pp, ISBN 978-88-470-0618-8

39. A. Quarteroni
Modellistica numerica per problemi differenziali (4a Ed.)
2008, XVI+560 pp, ISBN 978-88-470-0841-0

40. C. Canuto, A. Tabacco
Analisi Matematica II
2008, XVI+536 pp, ISBN 978-88-470-0873-1
2010, ristampa con modifiche

41. E. Salinelli, F. Tomarelli
    Modelli Dinamici Discreti (2a Ed.)
    2009, XIV+382 pp, ISBN 978-88-470-1075-8

42. S. Salsa, F.M.G. Vegni, A. Zaretti, P. Zunino
    Invito alle equazioni a derivate parziali
    2009, XIV+440 pp, ISBN 978-88-470-1179-3

43. S. Dulli, S. Furini, E. Peron
    Data mining
    2009, XIV+178 pp, ISBN 978-88-470-1162-5

44. A. Pascucci, W.J. Runggaldier
    Finanza Matematica
    2009, X+264 pp, ISBN 978-88-470-1441-1

45. S. Salsa
    Equazioni a derivate parziali – Metodi, modelli e applicazioni (2a Ed.)
    2010, XVI+614 pp, ISBN 978-88-470-1645-3

46. C. D'Angelo, A. Quarteroni
    Matematica Numerica – Esercizi, Laboratori e Progetti
    2010, VIII+374 pp, ISBN 978-88-470-1639-2

47. V. Moretti
    Teoria Spettrale e Meccanica Quantistica – Operatori in spazi di Hilbert
    2010, XVI+704 pp, ISBN 978-88-470-1610-1

48. C. Parenti, A. Parmeggiani
    Algebra lineare ed equazioni differenziali ordinarie
    2010, VIII+208 pp, ISBN 978-88-470-1787-0

49. B. Korte, J. Vygen
    Ottimizzazione Combinatoria. Teoria e Algoritmi
    2010, XVI+662 pp, ISBN 978-88-470-1522-7

50. D. Mundici
    Logica: Metodo Breve
    2011, XII+126 pp, ISBN 978-88-470-1883-9

51. E. Fortuna, R. Frigerio, R. Pardini
    Geometria proiettiva. Problemi risolti e richiami di teoria
    2011, VIII+274 pp, ISBN 978-88-470-1746-7

52. C. Presilla
    Elementi di Analisi Complessa. Funzioni di una variabile
    2011, XII+324 pp, ISBN 978-88-470-1829-7

53. L. Grippo, M. Sciandrone
    Metodi di ottimizzazione non vincolata
    2011, XIV+614 pp, ISBN 978-88-470-1793-1

54. M. Abate, F. Tovena
    Geometria Differenziale
    2011, XIV+466 pp, ISBN 978-88-470-1919-5

55. M. Abate, F. Tovena
    Curves and Surfaces
    2011, XIV+390 pp, ISBN 978-88-470-1940-9

56. A. Ambrosetti
    Appunti sulle equazioni differenziali ordinarie
    2011, X+114 pp, ISBN 978-88-470-2393-2

57. L. Formaggia, F. Saleri, A. Veneziani
    Solving Numerical PDEs: Problems, Applications, Exercises
    2011, X+434 pp, ISBN 978-88-470-2411-3

58. A. Machì
    Groups. An Introduction to Ideas and Methods of the Theory of Groups
    2011, XIV+372 pp, ISBN 978-88-470-2420-5

59. A. Pascucci, W.J. Runggaldier
    Financial Mathematics. Theory and Problems for Multi-period Models
    2011, X+288 pp, ISBN 978-88-470-2537-0

60. D. Mundici
    Logic: a Brief Course
    2012, XII+124 pp, ISBN 978-88-470-2360-4

61. A. Machì
    Algebra for Symbolic Computation
    2012, VIII+174 pp, ISBN 978-88-470-2396-3

62. A. Quarteroni, F. Saleri, P. Gervasio
    Calcolo Scientifico (5a ed.)
    2012, XVIII+450 pp, ISBN 978-88-470-2744-2

63. A. Quarteroni
Modellistica Numerica per Problemi Differenziali (5a ed.)
2012, XVIII+628 pp, ISBN 978-88-470-2747-3

64. V. Moretti
Spectral Theory and Quantum Mechanics
With an Introduction to the Algebraic Formulation
2013, XVI+728 pp, ISBN 978-88-470-2834-0

65. S. Salsa, F.M.G. Vegni, A. Zaretti, P. Zunino
A Primer on PDEs. Models, Methods, Simulations
2013, XIV+484 pp, ISBN 978-88-470-2861-6

66. V.I. Arnold
Real Algebraic Geometry
2013, X+110 pp, ISBN 978-3-642-36242-2

The online version of the books published in this series is available at SpringerLink.
For further information, please visit the following link:
http://www.springer.com/series/5418